Tea Environments and Plantation Culture

Arnab Dey examines the intersecting role of law, ecology, and agronomy in shaping the history of tea and its plantations in British East India. He suggests that looking afresh at the legal, environmental, and agroeconomic aspects of tea production illuminates covert, expedient, and often illegal administrative and commercial dealings that had an immediate and long-term human and environmental impact on the region. Critiquing this imperial commodity's advertised mandate of agrarian modernization in colonial India, Dey points to numerous tea pests, disease ecologies, felled forests, harsh working conditions, wage manipulation, and political resistance as examples of tea's unseemly legacy in the subcontinent. Dey draws together the plant and the plantation in highlighting the ironies of the tea economy and its consequences for the agrarian history of eastern India.

Arnab Dey is Assistant Professor of Modern Indian History at the State University of New York at Binghamton, and a visiting scholar at the South Asia Program, Mario Einaudi Center for International Studies, Cornell University. His research interests span the fields of labor, political economy, law, and environmental history. He has held fellowships from the Andrew W. Mellon Foundation, the Nicholson Center for British Studies at the University of Chicago, and the Rachel Carson Center for Environment and Society in Munich.

Tea Environments and Plantation Culture

Imperial Disarray in Eastern India

Arnab Dey
State University of New York at Binghamton

CAMBRIDGE
UNIVERSITY PRESS

University Printing House, Cambridge CB2 8BS, United Kingdom

One Liberty Plaza, 20th Floor, New York, NY 10006, USA

477 Williamstown Road, Port Melbourne, VIC 3207, Australia

314–321, 3rd Floor, Plot 3, Splendor Forum, Jasola District Centre, New Delhi – 110025, India

79 Anson Road, #06–04/06, Singapore 079906

Cambridge University Press is part of the University of Cambridge.

It furthers the University's mission by disseminating knowledge in the pursuit of education, learning, and research at the highest international levels of excellence.

www.cambridge.org
Information on this title: www.cambridge.org/9781108471305
DOI: 10.1017/9781108687034

© Arnab Dey 2018

This publication is in copyright. Subject to statutory exception and to the provisions of relevant collective licensing agreements, no reproduction of any part may take place without the written permission of Cambridge University Press.

First published 2018

Printed in the United Kingdom by TJ International Ltd. Padstow, Cornwall

A catalogue record for this publication is available from the British Library.

Library of Congress Cataloging-in-Publication Data
Names: Dey, Arnab, 1978– author.
Title: Tea environments and plantation culture : imperial disarray in eastern India / Arnab Dey, State University of New York, Binghamton.
Description: Cambridge, United Kingdom ; New York, NY : Cambridge University Press, 2018. | Includes bibliographical references and index.
Identifiers: LCCN 2018031740 | ISBN 9781108471305
Subjects: LCSH: Tea – India – History.
Classification: LCC SB272.I5 D49 2018 | DDC 633.7/20954–dc23
LC record available at https://lccn.loc.gov/2018031740

ISBN 978-1-108-47130-5 Hardback

Cambridge University Press has no responsibility for the persistence or accuracy of URLs for external or third-party internet websites referred to in this publication and does not guarantee that any content on such websites is, or will remain, accurate or appropriate.

For my parents, and in memory of my late sister

What can be a more melancholy sight in a savage country than to see our own countrymen greater savages than those, whom we call Cannibals and gross idolators. Of these poor ignorant islanders we may say Lord have mercy on them for they know not what they do – But what prayer must we offer for those who should know better and are every day doing worse

What then have we civilized done for the Savages? Will you say they were overwhelmed in vice before our arrival?

> True they had vices – such as Nature's growth
> But only the Barbarian's – We have both:
> The garden of civilization mixed
> With all the savage which man's fall had fixed
> ...
> The prayers of Abel linked to the deeds of Cain?
> Who such would see may from his lattice view
> The Old more degraded than the New

Journal of Thomas Machell, planter, India, 1824–62,
Mss. Eur. B 369/1, British Library, London

India is full of contrasts and, although Mahatma Gandhi, the anopheles mosquito and the planters' club may appear to have had little in common, they were, in fact, closely related.

John Rowntree, *A Chota Sahib: Memoirs of a Forest Officer*, 1981

Contents

List of Figures	*page* x
List of Maps	xi
List of Tables	xii
Acknowledgments	xiii
Note on Orthography	xviii
Introduction	1
1 Planting Empires	34
2 Agriculture or Manufacture?	49
3 Bugs in the Garden	77
4 Death in the Fields	97
5 Conservation or Commerce?	133
6 Plant and Politics	165
Conclusion	195
Bibliography	206
Index	224

Figures

0.1 An Assam plantation © Schlesinger Library, Radcliffe Institute, Harvard University *page* 10
0.2 "Tea as Progress," Item no. 13, *Tea Exhibition* © Hitesranjan Sanyal Memorial Collection, Centre for Studies in Social Sciences, Calcutta 16
1.1 Tea plant (*Camellia sinensis*): flowering stem. Watercolour © Wellcome Library 35
1.2 The Great Tea Race © The National Archives Image Library, United Kingdom 39
2.1 Cachar and Sylhet Tea Factory, Barak Valley, Assam, India, 1910, unattributed postcard © Mary Evans Picture Library 64
3.1 *Helopeltis theivora* Waterhouse (Tea mosquito) © Rison Thumboor, Wikimedia Commons, CC BY-SA 4.0 78
4.1 Charingia, Assam, India: kala-azar patients; a group of men, women, and children © Wellcome Library 111
4.2 A kalar-azar treatment center, Charingia, Assam, India: Indian patients are grouped outside a row of grass-roofed buildings, 1910 © Wellcome Library 113
4.3 "Capt. Robertson and Hospital Assistant throwing quinine into the mouths of loaded coolies." Sadiya, 1911–12 © The British Library Board, Photo 1083/34(163) 119
5.1 View looking over a tea plantation towards a river, Assam, 1860s © The British Library Board, Photo 682(111) 139
6.1 Group portrait of field laborers on a tea plantation, Assam, 1860s © The British Library Board, Photo 682(110) 166
6.2 The junction of the rivers [railway construction scene at junction of Ganges and Brahmaputra rivers near Goalundo], c. 1870 © The British Library Board, Photo 230(44) 171
6.3 Hoeing between the plants on an Indian tea plantation, 1911, postcard by unnamed artist © Mary Evans Picture Library 184

Maps

0.1 Assam and surrounding regions *page* 2
0.2 Principal tea-growing districts, Assam 8
3.1 Map showing tea mosquito bug attack on Ghazipore tea estate, 1908. The dark shaded portions show areas affected, with the darkest spots indicating severe damage © Adapted from C. B. Antram, *The Mosquito Blight of Tea: Investigations during 1908–09*, Indian Tea Association (Calcutta: The Catholic Orphan Press, 1910) 83

Tables

3.1 Statistics showing tea yields per acre, percentage increase or decrease, and variation from previous years *page* 88
4.1 Statement showing mortality on tea estates and number of unhealthy estates 126
4.2 Statement of adult death rates in a twenty-year period 128
6.1 Outturn of manufactured tea (in lbs.) per acre 180
6.2 Tea produced, and corresponding prices, for twelve months ending March 31 180

Acknowledgments

This project has had an interesting and somewhat unusual intellectual trajectory. While a very small part of it was (perhaps) embryonic in my doctoral dissertation, most of it was written afresh alongside a full-time academic appointment. There were numerous times when I felt this to be an impossible, if not foolish, undertaking. That it could be completed at all is due to the encouragement and help of many individuals and institutions across several continents. I owe them all my deep gratitude.

My dissertation mentors, Dipesh Chakrabarty and Fredrik Albritton Jonsson, deserve special mention in this list. They were the first to instill in me the confidence and ability to look beyond predictable historical answers and frameworks. This study owes its focus and inspiration to their scholarship, intellectual verve, and generosity during my time at the University of Chicago. I am also grateful to Muzaffar Alam, Ulrike Stark, Clinton Seely, Ralph Austen, Rochona Majumdar, Tanika Sarkar, Neeladri Bhattacharya, Kunal Chakrabarti, Samita Sen, Rana Behal, Prabhu Mohapatra, and Gautam Bhadra for comments and suggestions at various times during this project's doctoral tenure. The germ of this work took shape a very long time ago, and in a very different form, under the intellectually stimulating guidance of Udaya Kumar at the University of Delhi. I owe Udaya my sincere thanks for his continued support and friendship.

More recently, this project has benefitted from conversations with Prakash Kumar, Fa-ti Fan, Sudipta Sen, and Amitav Ghosh. I thank them for their time, suggestions, and constructive feedback. Prakash (da) and Fa-ti read several chapters and provided professional and personal support at several critical junctures. Two invited workshops in 2017, the first at the Agrarian Studies Program, Yale University, and the other at the Environment and Humanities Workshop, University of California at Davis, were immensely helpful in shaping and tying together the book's overarching arguments. At Yale, I am grateful to James Scott and Kalyanakrishnan Sivaramakrishnan (Shivi) for the invitation to present my work, and for their feedback. I also wish to thank Sahana Ghosh for her comments, and the other workshop participants for their suggestions and criticisms. At UC Davis, Louis Warren, Sudipta Sen, Kaleb Knoblauch, and Loren Michael Mortimer have my gratitude and thanks for stimulating conversations over two days.

Several segments of this project have also been presented at the annual American Historical Association conference, the annual conference of the American Society for Environmental History, the Association for Asian Studies conference, and at invited lectures at the Centre for Studies in Social Sciences (CSSSC), Kolkata; the department of Humanities and Social Sciences, Indian Institute of Technology, Guwahati; the Indian Institute of Science Education and Research, Pune, India; and at the Katholieke Universiteit, Leuven, Belgium. My thanks to all the conference participants and lecture attendees for their incisive comments and suggestions.

More than two full chapters were written during a six-month Carson Fellowship at the Rachel Carson Center (RCC), LMU, Munich in 2016. This wonderful institution for the environmental humanities was an ideal home for sustained writing and research. I thank its co-directors, Helmut Trischler and Christof Mauch, for giving me this opportunity and to my cohort of fellows – Salma, Lise, Ernst, Vimbai, Cindy, Saskia, Yan, Pey-Yi, Bob, Alan, and Paul for intellectual insights during weekly works-in-progress, conviviality, and epicurean solidarity. I owe Paul (Sutter) special acknowledgments for his detailed comments on my Introduction, and for his continued support for this project. I am also deeply grateful to RCC's staff, namely Arielle Helmick, Franz Langer, and Carmen Dines, for making my visit to Munich smooth and enjoyable.

The department of History at the State University of New York (SUNY) at Binghamton has been my intellectual and professional home since Fall 2012. As junior faculty, I could not have asked for a more collegial and supportive work environment. I thank my successive chairs, Nancy Appelbaum, Howard Brown, and Heather DeHaan, for facilitating leave requests, curriculum adjustments, and professional advice on many matters. I also wish to thank my colleagues Kent Schull, Nathanael Andrade, John Chaffee, Leigh Ann Wheeler, Fa-ti Fan, Jean Quataert, Elizabeth Casteen, Yi Wang, and Elisa Camiscioli for their support. I owe Elisa additional words of gratitude for her mentorship, conversations, and advice on matters academic and beyond. Keith Limbach, Colleen Marshall, and Kathy Fedorchak deserve special mention for their frequent help with administrative and departmental issues. My undergraduate and graduate students at Binghamton have been a delightful source of inspiration. I have learnt much from their curiosity and intellectual sharpness. My graduate seminar on "Empire, Bodies,

Power" in Fall 2017 was especially generative in thinking through methodological questions on spatial regulation and imperial "ruination" discussed in this work. I wish to thank Mariia Koskina, Zeynep Dursun, Bahattin Demir, and Şahika Karatepe for their interest and involvement in this class. My other graduate students, namely Ömer Topal, Lorena C. Duque, Zhixin Luo, Chulki Kim, and Erin Riggs have been wonderful interlocutors and co-researchers. The Office of the Dean, Harpur College of Arts and Sciences at Binghamton provided generous funds for research and conference travel, and assisted with leave paperwork. I am grateful to successive deans Anne McCall and Elizabeth S. Chilton for their commitment and support in this regard.

I can confidently say that this project would not have been possible, or would have been much delayed, without the institutional support and intellectual base of the South Asia Program (SAP), Mario Einaudi Center for International Studies at Cornell University. I wish to sincerely thank its former and present directors, Daniel Gold and Anne Blackburn, respectively, for extending me a formal affiliation since 2013. The participants of SAP's weekly colloquium, where portions of Chapters 3 and 4 were first presented, have my thanks for comments and suggestions. I am especially grateful to Anne, Durba Ghosh, Robert Travers, and Shelley Feldman for their friendship and intellectual generosity. Acknowledgments are also due to Bill Phelan, Daniel Bass, and the research staff of Olin and Mann libraries at Cornell for their help and support.

Many other librarians, archivists, and curators in India, the United Kingdom, and the United States have helped with this project. I wish to thank Dr. D. Sonowal, director of the Assam State Archives (ASA), Guwahati, for his support during my numerous visits, and to Haren Baishya for cheerfully attending to my many requests for documents. Baishya(da)'s knowledge of the ASA is now legendary. Sarat Chiring and Gogoida at the Dibrugarh District Collector (DC) record room, Sishuram Gogoi and Bhubenda at the DC Office, Jorhat, Assam, and the librarians at the Tocklai Tea Research Institute, Jorhat deserve my deepest gratitude for assisting me in the face of huge odds. I also wish to thank the librarian and Annex-I staff of the National Library of India, Kolkata for their help, and Abhijit Bhattacharya at the CSSSC, Kolkata for his assistance in procuring five archival images. Dr. Olive Geddes at the National Library of Scotland helped find a microfilm from their manuscript collection and Dr. Kevin Greenbank at the Centre for South Asian Studies, University of Cambridge gave generously of his time and knowledge in locating relevant source materials. I thank them both for their efforts. The archivists at the Asian and African Studies,

British Library, London; the Wellcome Library, London; the National Archives, Kew; the Center for Research Libraries, Chicago; the United States Department of Agriculture library, Beltsville, Maryland; the Schlesinger Library at the Radcliffe Institute for Advanced Study, Harvard University; and the Centre for Agriculture and Biosciences International (CABI), Oxfordshire all have my sincere thanks for their prompt and professional help with this project. Grace Vriezen at the University of Wisconsin-Madison Cartography Laboratory (UWCL) designed the two maps of Assam reproduced in this book. My thanks to Grace and UWCL's creative director, Tanya Buckingham, for their support and interest in this project.

Many friends, some within the academy and others beyond, have contributed to the successful completion of this project. In ways small, big, and incalculable, they have provided intellectual and logistical support, an emotional anchor, and the life stability needed for a book project's gestation. My archival work in Assam would not have been possible without the personal involvement of Jishnu Barua, Indian Administrative Service (IAS). Administrator, history buff, and friend all rolled into one, Jishnu (da) facilitated access to dusty and forgotten provincial *mahapeshkhanas* (record rooms), provided home comforts in Guwahati and Delhi, and commented engagingly on this developing tea story in colonial East India over many summers. I have learnt much from my conversations with Arupjyoti Saikia (Arupda) for over a decade now. Arupda's rigorous work on the history and society of Assam continues to provide inspiration and instruction in equal measure.

The entire Kanker family – at Kanker, Chhattisgarh, and in New Delhi – has provided a second home for close to two decades now. I cannot do justice to their contributions with a formal thank you such as this. Aditya Pratap Deo has been a personal and intellectual sounding board and inspiration since the first days of my undergraduate studies in Delhi. I owe Aditya my deep sense of gratitude for everything. Bikram Phookun has been important for as long; my thanks to him for advice, friendship, and generous hospitality in Delhi and Guwahati throughout the life of this project and before. Deeksha Bharadwaj in New Delhi, Pritom Phookun, and Archana *bou* in Oslo, Rajarshi and Utsa in Kolkata, Pratibha in Mumbai, and Arvind, Sreyoshi, Partha, and Dhruv and family in the United States have all provided support for this project and *joie de vivre* in ways that cannot be neatly categorized. I owe them all my sincere gratitude.

To be sure, no book project can be sustained by good wishes alone. I am grateful to have received outstanding financial and institutional support in the run-up to and throughout the duration of this research.

Two stints in the United Kingdom facilitated by The Nicholson Center for British Studies, University of Chicago provided primary materials that could only be used after the completion of my doctoral dissertation. I am also very thankful to the Harpur College Dean's office at SUNY Binghamton for the J. P. Mileur Faculty Development Fund Award, a book subvention grant, the Institute for Advanced Studies in the Humanities fellowship, and the Harpur Faculty Research Grant over these past six years. These have been immensely beneficial in the timely and planned completion of this project.

At Cambridge University Press, Lucy Rhymer went far beyond her role as senior acquisitions editor to provide encouragement, feedback, and support throughout the life of this project's publication. I owe Lucy a special word of thanks for her guidance and patience. Her colleagues, Lisa Carter and Ian McIver, expertly shepherded the book from production to print; my sincere gratitude to them both. I am much obliged to Lisa DeBoer for her work on the index. I also wish to thank the three anonymous reviewers of this manuscript for their helpful comments and suggestions. While I could not always follow their advice, this book reads in a more focused way due to their efforts. A longer version of Chapter 3 previously appeared in *Environment and History* 21, no. 4 (November 2015): 537–565, and Chapter 4 was published in a slightly modified form in *Modern Asian Studies* 52, no. 2 (2018): 645–82. I thank the editors of these journals for permission to reprint this material.

The most important I save for last. Mridu has been my companion and fellow researcher for as long as all this has mattered. She has weathered her own demanding academic life to provide support and understanding in immeasurable ways. For sharing her life and love with me, I want to say thank you. I am grateful to my in-laws in Jorhat for giving me the comforts of home and more during numerous research trips. The doddles make my world a wonderfully happy place; no word of thanks can repay it. Finally, it feels deeply inadequate to formally thank my parents, for what does any of this mean without them? I was fortunate to grow up in an academic household, yet one rooted in values of the everyday. For providing me with the best of everything, unwavering support, and love amidst life's toughest odds, I dedicate this book to my parents and in memory of my late sister.

Note on Orthography

The common practice of indicating "s" in the Assamese language with a velar fricative "x" in English has not been adhered to. Thus, I have retained "Assamese" (meaning both the language and the people of Assam) and not "Axomiya," as is usually noted with the fricative. Diacritics have not been used for vernacular terms and transliterations into English.

Place names and proper nouns (except as they appear in primary documents and place of first use) have been modernized.

Introduction

This book is about *Camellia sinensis var. assamica*, a tea plant grown in Assam in northeastern India. Celebrated worldwide for its body and flavor, it accounts for almost half of all Indian tea exports.[1] This study is also about the people and place that made it happen. A drive through the heart of the Assam tea country – a few hours east of Guwahati, the state capital – showcase these visually striking plantations. It is a picture of veritable calm. With lush green estates, neatly trimmed lawns, tea pickers in colorful attire, the occasional whirr of mechanical irrigators, and bracing winds of the surrounding hills, they are picture-postcard in natural beauty.

There is another side to the Assam tea story. It takes us back to British parleys in this northeastern frontier during the First Opium War to find an alternative base for this prized crop. It involves the military fiscalism of the expanding East India Company. The fortuitous find of this second Eden in the "wilds" of Assam sealed the shrub's fate. Vast swathes of her agrarian landscape – hitherto under direct or indirect possession of local peasant-cultivators, monastic orders, royal households, or autochthon hill "tribes" – were brought under the control of European tea speculators.[2] With Assam's formal annexation into the British Empire in 1826, other tea men, sanitarians, botanists, Company surgeons, and colonial administrators crowded into this newly acquired territory. The labor demands of this monoculture experiment were staggering, and form a crucial component of this other narrative. Except for some Kachari and Mising "tribals," initial attempts to lure or coerce local working hands into these plantations failed. With no dearth of cultivable land, peasants and Assamese agriculturists were unwilling to trade the freedom of homestead cultivation for the restrictive work environment on

---

[1] Vide www.teaboard.gov.in/pdf/bulletin/Estimated_production_for_Apr_2017.pdf (accessed June 2, 2017).
[2] See Indrani Chatterjee, *Forgotten Friends: Monks, Marriages, and Memoirs of Northeast India* (New Delhi: Oxford University Press, 2013), especially chapter 5.

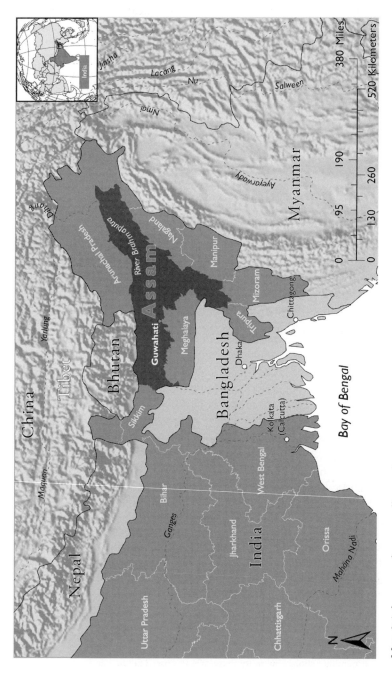

Map 0.1 Assam and surrounding regions

these estates. Faced with an uncooperative local labor market, and intense colonial competition, Assam turned to indentured recruitment after 1865. Millions of men, women, and children were brought in from central and south-central India to these emergent plantations. Migrants of economic necessity, or often of recruiter guile, they faced inhospitable terrain, unhygienic conditions on these plantations, managerial oversight, and harsh working conditions in this unknown land. Unthinkable labor mortality during the colonial period and beyond followed.[3] This side of the story is also about continued planter brutality, and managerial violence that lent enduring notoriety to the Assam plantation system. Periodic flare-ups in the form of worker protests, riots, desertions, and walkouts were not uncommon. Meanwhile, the socio-economic opportunities created by this tea enterprise drew in middle-ranking estate functionaries, small-time creditors, and share-croppers from Bengal and elsewhere that, in turn, definitively changed Assam ethnic demography. Tea, of course, has always been an export product and added very little to the region's economic coffers during the colonial period, and thereafter. This history of ethnic immigration, and extractive commodity capitalism snowballed into regional disquiet in the decades following India's independence – first with a student-led movement for greater local autonomy, and then into a full-blown armed insurgency against "neglect" by the postcolonial Indian State.[4]

This is the social history of the Assam plantations that has been repeatedly told. Indentured tea labor, of course, is part of another global narrative: the intra-colonial and transoceanic relocation of contract work in the aftermath of abolition. Aiding the rising production demand of plantation cash crops, this human traffic ranged across the imperial meridian – from India to Mauritius, Fiji, British Guiana, Surinam, French Guiana, Martinique, Guadeloupe, Réunion, and Malaya among others. Within the subcontinent, this era of indentured overseas emigration saw stiff competition from the coffee, tea, indigo and jute growing regions of eastern and southern India, the coal mines of Bihar and the textile manufacturing industries of the Bombay Presidency. There has been much scholarly focus on the nature and form of these labor migrations, their socio-demographic and economic push-factors, demand and

---

[3] Despite their shortcomings, consider two recent journalistic findings on the Assam plantations that sum up the contemporary legacy of their colonial history: one by *The New York Times* titled "Hopes, and Homes, Crumbling on Indian Tea Plantations," (February 13, 2014) and the other by the BBC, "The Bitter Story behind the UK's National Drink" (September 8, 2015).

[4] Indeed, this capital-intensive enterprise was almost wholly an "alien" import; besides land, all other factors of production were brought in from other parts of India or metropolitan Britain; the scholarship on these issues is discussed below.

supply patterns, and impact on resettlement and (de) peasantization in the home and host countries.[5] That this traffic in "voluntary" and "free" labor movement in the Old World never really emerged out of slavery's shadow in the New is well established. As a British Consul in Paramaribo expressed in 1884, "the Surinam planters ... found in the meek Hindu a ready substitution for the negro slave he had lost."[6] Whether or not such migration stemmed from volition or coercion, historians on both sides of the debate have drawn our attention to the structural forms of exploitation, conditions of work and travel, the cycles of debt-bondage, and subsistence wages that underpinned these sites of European agro-business. It is hard to miss Hugh Tinker's foundational influence in these explorations of labor life in an era of indentured contract.[7]

As far as the Assam plantations are concerned, three broad approaches dominate in the existing scholarship: its working-class history, regional political fallout, and ethno-social impacts. For the first, labor historians have long remarked on an overbearing work regime, miserable conditions to and on these gardens, brutal planters, and systemic managerial license.[8] Indeed, a recent work reiterates that the political economic hallmark of the Assam system was a combination of coercive power

---

[5] The literature on indentured migration, or on plantation systems overall, is vast; see the helpful, though dated, bibliography compiled by Edgar T. Thompson, *The Plantation: A Bibliography*, Social Science Monographs IV (Washington, DC: Pan American Union, 1957); also see P. C. Emmer, ed., *Colonialism and Migration; Indentured Labour Before and After Slavery* (The Netherlands: Martinus Nijhoff, 1986).

[6] Emmer, *Colonialism and Migration*, p. 187.

[7] Hugh Tinker, *A New System of Slavery: The Export of Indian Labor Overseas, 1830–1920* (London: Institute of Race Relations, 1974). Also see Philip Corrigan, "Feudal Relics or Capitalist Monuments? Notes on the Sociology of Unfree Labor," *Sociology* 11(3) (1977): 435–463; Robert Miles, *Capitalism and Unfree Labor: Anomaly or Necessity?* (London: Tavistock Publications, 1987).

[8] These include Rajani Kanta Das, *Plantation Labor in India* (Calcutta: Prabasi Press, 1931); Ranajit Das Gupta, *Labor and Working Class in Eastern India: Studies in Colonial History* (Calcutta and New Delhi: K. P. Bagchi & Company, 1994); Sharit Bhowmik, *Class Formation in the Plantation System* (New Delhi: People's Publishing House, 1981); Sanat Bose, *Capital and Labor in the Indian Tea Industry* (Bombay: All India Trade Union Congress, 1954); Muhammad Abu B. Siddique, *Evolution of Land Grants and Labor Policy of Government: The Growth of the Tea Industry in Assam 1834–1940* (New Delhi: South Asian Publishers, 1990); J. C. Jha, *Aspects of Indentured Inland Emigration to North-East India 1859–1918* (New Delhi: Indus Publishing Company, 1996); Rana P. Behal and Prabhu P. Mohapatra, "Tea and Money Versus Human Life: The Rise and Fall of the Indenture System in the Assam Tea Plantations 1840–1908," *Journal of Peasant Studies* 19(3) (1992): 142–172; Rana Partap Behal, "Forms of Labor Protests in the Assam Valley Tea Plantations, 1900–1947," *Occasional Papers on History and Society* (New Delhi: Nehru Memorial Museum and Library, 1997); Behal, "Power Structure, Discipline and Labor in Assam Tea Plantations Under Colonial Rule," *International Review of Social History* 51 Special Supplement (2006): 143–172; Samita Sen, "Commercial Recruiting and Informal Intermediation: Debate over the Sardari System in Assam Tea Plantations, 1860–1900," *Modern Asian Studies* 44.1 (2010): 3–28; see also,

structure, unregulated immigration, and extra-legal planter authority all rolled into one.[9] Secondly, as the region's primary socio-economic driver, these plantations have been noted for their importance in forming Assamese political consciousness.[10] Lastly, they have been examined for their role in creating ethnic and sub-national claims to "homeland" and "otherness" in the region.[11] The enduring and influential epithet of a "Planter's Raj" in the second approach argued that the planters' hold over civic, socio-economic, and political power in the province throughout the long nineteenth century stymied – but simultaneously created – the conditions for the eventual rise of a regional bourgeoisie who, with support from the Indian National Congress, reclaimed the political mantle in Assam in the decades preceding independence.[12] The material imperatives of "improvement" and "progress" of this imperial tea regime in the third assessment – partially put to use by the educated Assamese middle-class "elites," by those taking advantage of its opportunities from Bengal, by itinerant graziers and herdsmen from Nepal, and by the relocated indentured émigrés from "outside" – are credited with creating the historical and social conditions of "exclusionary" and sectarian Assamese solidarities, ethnic divisions, and supra-national "homeland" demands in postcolonial northeastern India.[13]

Agronomy, Ecology, and Plantation "Science"

But were the Assam plantations self-serving economic or social structures alone? By 1905, its total production area had swelled to more than

---

Bodhisatwa Kar, *Framing Assam: Plantation Capital, Metropolitan Knowledge and a Regime of Identities, 1790s–1930s*, unpublished PhD dissertation (New Delhi: Jawaharlal Nehru University, 2007); and Nitin Varma, "Coolie Acts and the Acting Coolies: Coolie, Planter and State in the Late Nineteenth and Early Twentieth Century Colonial Tea Plantations of Assam," *Social Scientist* 33(5/6) (2005): 49–72. Also see Dwarkanath Ganguly, *Slavery in British Dominion*, ed. Siris Kumar Kunda (Calcutta: Jijnasa Publications, 1972); Sir J. H. S. Cotton, *Indian and Home Memories* (London: T. Fisher Unwin, 1911); Mrs. Emma Williams, "Letter Regarding Abuses on the Tea Plantations of Assam," IOR/L/PJ/6/749, March 24, 1906, British Library London; Report from Aborigines Protection Society on "Treatment of Tea Labourers in Assam," IOR/L/PJ/6/193, January 17, 1887; Revered C. Dowding, "Letters and Pamphlets on the Illegal Arrest of Run-Away Tea-Garden Coolies in Assam," IOR/L/PJ/6/832, October 22, 1907, and the numerous House of Commons Parliamentary papers on the topic.
[9] See Rana Partap Behal, *One Hundred Years of Servitude: Political Economy of Tea Plantations in Colonial Assam* (New Delhi: Tulika Books, 2014).
[10] See Amalendu Guha, *Planter Raj to Swaraj: Freedom Struggle and Electoral Politics in Assam, 1826–1947* (New Delhi: ICHR, 1977, rpt. 2006).
[11] See Jayeeta Sharma, *Empire's Garden: Assam and the Making of India* (Durham, NC and London: Duke University Press, 2011).
[12] See Guha, *Planter Raj to Swaraj*, especially chapters 2, 3 and 4.
[13] See Sharma, *Empire's Garden*, especially part II.

338,000 acres.[14] More than 16 legislative Acts controlled one facet of its operation or another. By this same year, the labor of more than 2 million men, women, and children[15] had gone into producing this commodity. Labor mortality rate stood at a staggering 53.2 per thousand working adults.[16] Two imperial Labor Enquiry Commissions had visited these gardens by the second decades of the twentieth century, and had submitted their findings. Colonial administrators and nationalist leaders (including M. K. Gandhi) had spent their energies discussing various aspects of these pioneer estates. "Rival" resource stakeholders, namely the Indian Forest department had rubbed shoulders – often uncomfortably, but also out of necessity – with zealous guardians of this tea enterprise. A vigorous traffic in scientific opinion, field experiments, and personnel moved between Assam, Calcutta, Java, Ceylon (now Sri Lanka), London, and beyond. All this while, more than eight species of plant bugs and pests parasitically fed on the tea micro-climate and ravaged crop yields, flavor, and profits.

How are these histories connected, if at all? What agronomic, legal, and economic logics bring together these disparate features of the industry? What does tea's built environment have to do with labor protests, and conditions of worker impoverishment and morbidity? What role did nonhuman agents play in this monoculture economy? How did scientific discourse, agrarian ideology, and plantation practice come together?

This book answers many of these unexplored and seemingly unrelated questions. In what follows, it provides an agro-ecological history of tea production in colonial eastern India over a hundred-year period and beyond. In contrast to existing debates, I argue that a syncretic look at the legal, environmental, and agronomic aspects of tea production help us better understand *why* human and natural reordering in the region had overlapping, and invisible agendas. By using tea's self-avowed mandate of agrarian reform and modernization as its overall base, this book demonstrates that the enterprise was essentially a "knowledge economy," a congeries of ideological, scientific, and legal interests that did not always converge, or control opinion and outcome. That this disorderly house – the sum and part of what I heuristically and analytically call "disarray" – ultimately manifested itself in harsh working conditions, tea pests, disease

---

[14] The figures are for 1901, cited in Amalendu Guha, *Planter Raj to Swaraj*, p. 28. See also his "A Big Push without a Take-Off: A Case Study of Assam 1871–1901," *Indian Economic and Social History Review*, 5 (September 1968): 202–204.
[15] The figures are for 1905, see *The Report on Labor Immigration for the Province of Assam for the Year 1906* (Shillong: Assam Secretariat Press, 1906).
[16] The figures are for 1900; see Guha, *Planter Raj to Swaraj*, 2nd ed., p. 30.

Introduction 7

environments, labor mortality, wage manipulation, felled forests, and lawlessness is crucial to this book's overarching departure from existing narratives. The plant and the plantation are thus brought together to rethink this classic opposition between labor and capital.[17]

Like its peers elsewhere, the Assam plantations were also fueled by an elaborate environmental imagination and blueprint.[18] If the commodity was at the center of these plans, these imaginations also took shape in the midst of racial and social hierarchies, ecological improvisations, unintended consequences, agrarian practices, and complex rearrangements of labor and landscape. This book highlights those overlaps. Indeed, by some measure, tea was a demanding cash crop. As with similar tropical products – tobacco, for instance – crop success and capital investment were not directly proportional. Its ecological context was as important, if not more so, in the making and unmaking of production goals and plantation practice. If tea's natural setting has been underplayed and ignored in present accounts, this book shows that the physical environment played a variety of roles in this commodity history. Instrumentally, of course, climate, soil, moisture, rainfall, and overall weather patterns were inextricably linked to company fortunes. But beyond this, nature was an ideological battleground where matters of "wilderness" (anthropogenic or botanical), imperial power, agrarian "improvement," horticultural authority, and even fiscal imposts were fought and tested. If this human–nature link connects this work to a staple of environmental history, it is not the debate between pristineness and degradation that primarily concerns us here.[19] Indeed, as I discuss below, the

---

[17] On this perspective, see James L. A. Webb, Jr., *Tropical Pioneers: Human Agency and Ecological Change in the Highlands of Sri Lanka, 1800–1900* (New Delhi: Oxford University Press, 2002); James S. Duncan, *In the Shadows of the Tropics: Climate, Race and Biopower in Nineteenth Century Ceylon* (London: Ashgate Publishing Co., 2007); Corey Ross, *Ecology and Power in the Age of Empire: Europe and the Transformation of the Tropical World* (Oxford: Oxford University Press, 2017); Ann Laura Stoler, *Capitalism and Confrontation in Sumatra's Plantation Belt, 1870–1979* (New Haven, CT and London: Yale University Press, 1985); and Lynn A. Nelson, *Pharsalia: An Environmental Biography of a Southern Plantation, 1780–1880* (Athens, OH: University of Georgia Press, 2007).

[18] See the interesting collection of articles in Frank Uekötter, ed. *Comparing Apples, Oranges, and Cotton: Environmental Histories of the Global Plantation* (Frankfurt and New York, NY: Campus Verlag, 2014).

[19] See, for instance, Donald Worster, *Nature's Economy: A History of Ecological Ideas* (Cambridge: Cambridge University Press, 1988); Gregg Mitman, *The State of Nature: Ecology, Community, and American Social Thought, 1900–1950* (Chicago, IL: University of Chicago Press, 1992); and William Cronon, *Nature's Metropolis: Chicago and the Great West* (New York, NY: W. W. Norton & Co., 1991); their respective bibliographies provide a good source for this rich historiography. Cronon's distinction between "first" (pre-human) and "second" (post-human intervention) nature in *Nature's Metropolis*, p. 56 is instructive here.

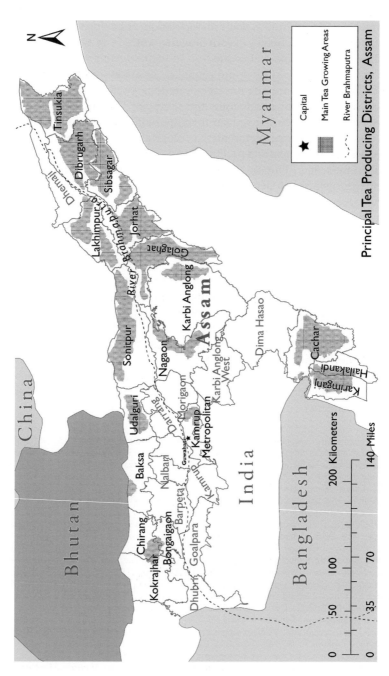

Map 0.2 Principal tea-growing districts, Assam

Introduction 9

Enlightenment parable of Edenic recovery underwent exceptional extramarket and extra-legal tweaks in British east India. In this study, nature is rather used as an ecosystem context to understand its expedient use and abuse in human and landscape transformations.[20] For instance, the agroecology that sustained tea plant growth also gave rise to blights and pests. Embankment works that irrigated these plantations, and paddy cultivation that fed sustenance-wage workers produced malaria and black-fever that, in turn, led to rampant labor ill-health and death. Aspects of tea cultivation that demanded field rigor, namely hoeing and plucking were utilized to prop up an illegal task-based labor wage system that favored bodily capacity over guaranteed monthly pay. Forests that provided necessary shade to tea saplings and provided wood for tea-boxes also sustained virulent malarial parasites that killed vast number of workers. Nature had many functions in this capital-intensive economy; it was the lynchpin between crop and capital. Frank Uekötter's argument is poignant in this context: that in its hegemonic ability and desire to condition nature as well as society, "plantation systems are akin to totalitarian states – matters of life and death for entire economies and regions."[21]

This work is therefore not about *a* specific commodity or *a* specific place. It is a plantation history first and foremost, and seeks to situate Assam within its broader Asia-Pacific and Atlantic contexts. But it does not call for this comparison through the primacy of any one approach – cultural, biological, or Marxist.[22] If labor is still at

---

[20] On this perspective, I draw inspiration from John Soluri's, *Banana Cultures: Agriculture, Consumption, and Environmental Change in Honduras and the United States* (Austin, TX: University of Texas Press, 2005); Sven Beckert, *Empire of Cotton: A Global History* (New York, NY: Vintage, 2014); Sidney W. Mintz, *Sweetness and Power: The Place of Sugar in Modern History* (New York, NY and London: Viking, 1985); Timothy Mitchell, *Rule of Experts: Egypt, Techno-Politics, Modernity* (Berkeley, CA and London: University of California Press, 2002); J. R. McNeill, *Mosquito Empires: Ecology and War in the Greater Caribbean, 1620–1914* (Cambridge: Cambridge University Press, 2010); Paul S. Sutter, "Nature's Agents or Agents of Empire? Entomological Workers and Environmental Change during the construction of the Panama Canal," *Isis*, Vol. 98, No. 4 (December 2007), 724–754; Richard White, *The Organic Machine: The Remaking of the Columbia River* (New York, NY: Hill and Wang, 2005); T. H. Breen, *Tobacco Culture: The Mentality of the Great Tidewater Planters on the Eve of Revolution* (Princeton, NJ: Princeton University Press, 1985); Richard S. Dunn, *Sugar and Slaves: The Rise of the Planter Class in the English West Indies, 1624–1713* (Chapel Hill, NC: University of North Carolina Press, 1972); Ian Tyrrell, *True Garden of the Gods: Californian-Australian Environmental Reform, 1860–1930* (Berkeley, CA: University of California Press, 1999), Uekötter, ed., *Comparing Apples, Oranges, and Cotton*, and Corey Ross, *Ecology and Power in the Age of Empire*.
[21] See Uekötter, ed., *Comparing Apples, Oranges, and Cotton*, p. 18.
[22] Consider, for instance, that in comparing the plantation economies of Assam and British West Indies, Prabhu P. Mohapatra focuses exclusively on the penal provisions of

10   Tea Environments and Plantation Culture

Figure 0.1 An Assam plantation © Schlesinger Library, Radcliffe Institute, Harvard University

the center of this book, an agroecological perspective de-centers attachment to categories such as proletariat and peasant, feudal and capitalist in understanding confrontation, or the relationship between crop and cultivator in extractive production. Entomological science, fertilizers, soil management, pathogen environments, and botanical manipulation contribute to, and are components of, planter violence and legal control of land and labor. As John Soluri suggests in his study of Honduran banana production, "attempts to draw well-defined borders between natural spaces and cultural spaces run the peril of ignoring all-important interactions between fields, forests, and waterways; and between cultivated, wild, and hybrid organisms."[23]

indentured contract and their mechanisms of enforcement; surely, the legal extraction of labor-power, and the resistance it evoked have other material and historical drivers in these two cases; see Prabhu P. Mohapatra, "Assam and West Indies, 1860–1920: Immobilizing Plantation Labor," in Douglas Hay and Paul Craven, *Masters, Servants, and Magistrates in Britain and the Empire, 1562–1955* (Chapel Hill, NC and London: The University of North Carolina Press, 2004), pp. 455–480.

[23] Soluri, *Banana Cultures: Agriculture, Consumption, and Environmental Change in Honduras and the United States*, p. 5.

Introduction   11

Secondly, this study adds to the debate on "imperial science" as a mélange of metropolitan and local knowledge.[24] But it is not just the dialogic and hybrid production of colonial modernity that is of interest here.[25] Rather, certain aspects of the tea story (epidemiology and pest ecology, for example) show that Western scientific "rationality" was a non-starter to begin with; that it was no more a tool of modernization than an unsure foray into what the industry little-understood, or in some cases, expediently forsook in the service of profits.[26] If scientific expertise was Empire's claim to power, I demonstrate that the gap between esoteric laboratory knowledge and field experience, between ideological beliefs and on-the-ground decisions, between clinical planning and actual outcome was vast and noteworthy. The situation was not too dissimilar in the nineteenth-century Ceylon coffee plantations. T. J. Barron notes that when it came to deciding between agrarian scientific knowledge and practical necessity, nineteenth century coffee planters in the island invariably chose the latter.[27] Barron argues that political contacts, discriminatory land and labor policies, favorable taxation rates, and preferential market access – rather than ostensibly "superior" Western scientific technology – allowed Ceylon planters to succeed. He notes: "science was thrown to the winds if thereby profitability could be increased."[28] The following chapters (especially 2, 4, and 5) show that their counterparts in eastern India were not too far off in this matter.

Indeed, in keeping nature tertiary to their concerns, Indian labor historians have unconsciously ascribed to plantation "science" a hegemony

---

[24] See the helpful analytic distinction between "imperial science" and the "science of Empire" in David Gilmartin's "Scientific Empire and Imperial Science: Colonialism and Irrigation Technology in the Indus Basin," *The Journal of Asian Studies*, 53.4 (November 1994): 1127–1149.

[25] See, for instance, Gyan Prakash, *Another Reason: Science and the Imagination of Modern India* (Princeton, NJ: Princeton University Press, 1999); also, Deepak Kumar, *Science and the Raj, 1857–1905* (New Delhi: Oxford University Press, 1995), Kapil Raj, *Relocating Modern Science: Circulation and the Construction of Knowledge in South Asia and Europe, 1650–1900* (Basingstoke: Palgrave, 2007), and David Arnold, *Science, Technology and Medicine in Colonial India* (Cambridge: Cambridge University Press, 2000). Also see the discussion of indigo knowledge and science in Prakash Kumar, *Indigo Plantations and Science in Colonial India* (Cambridge and New York, NY: Cambridge University Press, 2012).

[26] The thorny relationship between scientific "expertise" and imperial exploration has been much commented upon. For a stimulating discussion of this overlap in the African context, see Helen Tilley, *Africa as a Living Laboratory: Empire, Development, and the Problem of Scientific Knowledge, 1870–1950* (Chicago, IL: University of Chicago Press, 2011).

[27] See T. J. Barron, "Science and the Nineteenth-Century Ceylon Coffee Planters," *The Journal of Imperial and Commonwealth History* 16, 1 (1987): 5–23.

[28] Ibid., p. 7.

that it neither possessed nor wholly controlled. To that end, the following chapters show how imperial drives of conquest and desires for wealth changed ecologies and how those environments, in turn, modified plantation operations in nonlinear and coevolving ways. The parallels with Ceylon are again instructive. As with tea in India, the coffee plantation (and later cinchona and tea) mania in the island territory during the late 1830s led to large-scale ecological imbalance and social transformations.[29] Coffee pests and bugs, cattle disease, conflicts over pasture and water with local Kandyan villagers, deforestation, and worker mortality due to malaria and cholera provide a poignant foil to similar patterns of environmental transformation in northeastern India during the same period. The impact of Bordeaux spraying on banana plants to control the Sigatoka (*Mycosphaerella musicola*) fungus in the Honduran Sula valley in the 1930s in terms of messy residues, reduced yields, and worker respiratory problems is yet another example.[30] These, and numerous other similar instances, show that the inherent Whiggish "rationality" of these large-scale monocultures was unsustainable from the very start.[31] In Timothy Mitchell's words, the "ideality of human intentions and purposes" and the "object world upon which these work" are never one-directional, predictable, or separable.[32] In exploring these overlaps in the making of Assam tea, therefore, this study resists putting nature on one side, and human calculation and expertise on the other.[33]

---

[29] See James L. A. Webb, Jr., *Tropical Pioneers: Human Agency and Ecological Change in the Highlands of Sri Lanka, 1800–1900*, especially chapters 3, 4, and 5; also, James S. Duncan, *In the Shadows of the Tropics: Climate, Race and Biopower in Nineteenth Century Ceylon*.

[30] See Soluri, *Banana Cultures*, especially chapter 4, pp. 104–127.

[31] Paul Sutter's analysis of the role of entomological workers during the construction of the Panama Canal provides yet another historical example. Sutter mentions: "my argument is not that scientists give us an unmediated access to material environmental agency – that they are, in a sense, nature's agents. Nor do I intend to imply that they are the only group in the imperial field who work across this gap between the material environment and idealized nature. Rather, my aim is to suggest that material environmental influence can be seen quite clearly at the points of tension between ideological predisposition and empirical observation," in "Nature's Agents or Agents of Empire?" p. 729.

[32] See *Rule of Experts: Egypt, Techno-Politics, Modernity*, especially part I.

[33] Mitchell, *Rule of Experts*, p. 36; for a fascinating study of the importance and agency of the cotton boll weevil, the Vedalia beetle, the corn borer, the San Jose scale and other pests in the history of American agricultural innovation, see Alan L. Olmstead and Paul W. Rhode, *Creating Abundance: Biological Innovation and American Agricultural Development* (Cambridge and New York, NY: Cambridge University Press, 2008); Olmstead and Rhode demonstrate that biological innovation and mechanical technologies did not follow each other chronologically in American agriculture but that in the two centuries before World War II, steady (but non-institutionalized) advancement in biological innovation in crop and livestock sectors increased both land and labor productivity ... that "American agricultural development was far more dynamic than generally portrayed," in *Creating Abundance*, p. 16.

Introduction   13

Doing so also challenges the three historiographical approaches indicated earlier. As far as the question of labor is concerned, I contend that worker exploitation in Assam went far beyond methods of bodily regulation and structures of recruitment. Indeed, the expansive Marxist historiography on this topic has underplayed and ignored the importance of ideological, epidemiological, and agronomic strategies and manipulations in the making of worker impoverishment and unrest in the region. This book shows that it was not just through direct physical violence or predictable fiscal maneuvers that the repressive relationship between master and men were forged in the Assam gardens. It was also linked to how pathogens, plants, legality, and landscapes were managed in expedient and unseen ways. If the labor body eventually bore the burden of these economic and para-economic managerial tactics, I argue that examining the multiple networks of those tactics are vitally important if we are to resist representing – and thereby privileging – only those aspects of coercion that are visible and visceral. In looking at labor's role, use and abuse in the Assam gardens as co-opted and co-related with features of the agro-environmental and the medico-legal, therefore, this book reorients the study of labor (and labor exploitation) in tropical plantations beyond its deterministic material-relations-of-production focus.

Similarly, if the reach of the so-called planter's raj has been rightly critiqued,[34] this book demonstrates that that Raj was established not just through extra-legal and extra-economic manipulations, but also by extra-human and agro-economic means. As it is, the dealings between the planting community and the colonial government during the period under study show that the parameters of *what* constituted "political" authority, *who* had it, and *how* much were continually under crisis in Assam. As with the Ceylon plantations, there were multiple sites of power and de-facto sovereignties within the Assam estates that dictated policy and decision-making.[35] Finally, analyzing the social impact of the tea enterprise in terms of its ethnic, regional, or national fallout does not tell us Improvement's story inside these plantations. For projects of "progress" and "modernity" were constituted and confounded by multiple contexts – human and nonhuman – in these gardens. My overarching quibble with these last two approaches, however, is that, to some degree

---

[34] See Hiren Gohain, "Politics of a Plantation Economy," Review of Amalendu Guha, *Planter's Raj to Swaraj* in *Economic and Political Weekly* Vol. 13, No. 13 (April 1 1978): 579–580.

[35] See James S. Duncan, *In the Shadows of the Tropics: Climate, Race and Biopower in Nineteenth Century Ceylon*, especially chapters 4, 5, and 6. I explore conflicts over authority and jurisdiction between the Indian Tea Association, Calcutta agency houses, on-site planters, and the colonial government in chapters 3, 4, 5 and 6.

or another, both use the Assam plantations as a heuristic, historical springboard to launch analyses of political or ethno-social conditions in the region. A large part of their methodology therefore focuses on the industry's impact *without* than on its praxis *within*. This book takes the latter approach to connect unexplored parts (the planter, the anopheles, and Gandhi as one of our epigraph author puts it) of this commodity's history in northeastern India. It finally reminds readers that tea's distinctive socio-environmental story cannot be told by keeping the plant on the one side, and the plantation on the other.

Tea *as* "Improvement"

the benefits which the tea industry has conferred on the Province have been many and great. The land most suitable for tea is not adapted to the cultivation of rice, and the greater part of it would still be hidden in dense jungles if it had not been cleared by the tea planters ... a great impetus has also been given to trade, and new markets have been opened in all parts of the Province. The existence of the tea industry has been a potent factor in the improvement of communication by rail, river and road.[36]

Assam's census commissioner and noted civil servant, Sir Edward Gait, minced no words about tea's purported role in the region. Long before Gait, however, a blueprint for European agrarian colonization – centered on the plant – had already been laid for the province. In this pantheon of "improvers," Captain Jenkins's name stands pre-eminent. Taking over from David Scott as the Agent to the Governor-General of the North-East Frontier in 1831, Jenkins used the Charter Act of 1833 to promote his vision of land, taxation, communication, and agrarian "progress" in Assam. Harping on the new-found promise of tea, Jenkins argued that only by inducing European speculators to the vast "wastes" of the region could the then "unpromising state of the country" be remedied. To that end, he laid out his plan to the visiting judge of the Calcutta Sadr Court, A. J. Moffatt Mills in 1853 that was to inaugurate the scramble for tea. While details of this plan are discussed in Chapter 5, some of Jenkins's ideas are worth examining here. In essence, Jenkins proposed a comprehensive system of management that would encourage "free intercourse" in trade with other provinces, open up communication through roads and steam, induce the Assamese to more permanently settled landholdings (then in operation only in the western Goalpara and the southern Sylhet districts), and increase revenue by recalibrating taxation on all heads except *rupit* (or wet-rice

---

[36] Sir Edward Gait, *A History of Assam* (Calcutta: Thacker, Spink, rpt. 1967), p. 413.

Introduction 15

cultivating) lands. Indeed, by recommending a tax on *bari* (or homestead) lands earlier in 1836, Jenkins had evinced hope that "these wastes and haunts of wild beasts" would transform into "fruitful fields of sugarcane, mustard, mulberry, lac, tobacco and vegetables."[37] To be sure, these were not European visions alone. Plans for agrarian "reform" centered on tea resonated with a select group of English-educated local middle-class who echoed its purported agenda.[38] As historian Jayeeta Sharma observes:

> A range of interlocutors, from British bio-prospectors to American missionaries to Assamese gentry, extolled the Edenic transformations under way, of a jungle into a garden. They conjured up a future ordered landscape of export-producing tea plantations, a stark contrast to the partially cultivated and imperfectly commercialized state of Nature that they saw in the present.[39]

Among these locals was a young civil servant and publicist Anandaram Dhekial Phukan who expressed confidence in the "beneficial" aspects of English rule in Assam, and the avenues it opened up for indigenous entrepreneurship, prosperity, and progress. In a lengthy petition to the visiting judge Mills, Dhekial Phukan entreated the British Government to "effect at once a complete [...] change in the agricultural prospects of the country" to render her produce tenfold more abundant.[40] Such was Jenkins's charm that in the opinion of noted Assamese historian S. K. Bhuyan, he was nothing short of "a worthy successor of the Assamese *Swargadeos* ... (literally, the Ahom god-kings)."[41]

Dhekial Phukan's hopes were to be dashed, for Jenkins soon changed his tone and ambitions. Already by June 6, 1853, the directors of the Assam Company had petitioned Governor-General Dalhousie that their sedulous efforts to transform uninhabited jungles into "smiling" tea gardens be afforded all legitimate degrees of aid.[42] A few days later, on June 22, Major H. Vetch, Deputy Commissioner of Assam urged Judge Mills that "even greater benefits" and "bounties" be laid out for

---

[37] H. K. Barpujari, *Assam: In the Days of the Company*, p. 215.
[38] For a study of the Assamese middle class, see Hiren Gohain, "Origins of the Assamese Middle Class," *Social Scientist*, Vol. 2, No.1 (August 1973): 11–26.
[39] Sharma, *Empire's Garden*, p. 3.
[40] See "Observations on the Administration of the Province of Assam, by Baboo Anundaram Dakeal Phookun," in A. J. Moffatt Mills, Esq., *Report on the Province of Assam* (Calcutta: Gazette Office, 1854), appendix J, p. xxxviii.
[41] S. K. Bhuyan, *Early British Relations with Assam: A Study of the Original Sources and Records Elucidating the History of Assam from the Period of its First Contact with the Honourable East India Company to the Transfer of the Company's Territories to the Crown in 1858* (Shillong: Assam Govt. Press, 1949), p. 31.
[42] Memorial of the Assam Tea Company to the Marquis of Dalhousie, Governor General of India, 6 June 1853, reproduced in A. J. Mills's *Report*, appendix E, pp. xix–xxi.

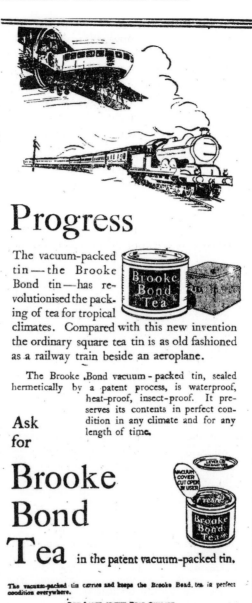

Figure 0.2 "Tea as Progress," Item no. 13, *Tea Exhibition* © Hitesranjan Sanyal Memorial Collection, Centre for Studies in Social Sciences, Calcutta

Introduction

prospective capitalists with generous land grants and by remitting their purchase money for three years and upwards.[43] As the Tea Committee's recommendations took off, and Assam Company's operations began, Jenkins's consolidated his peers' ideas for "accelerating the progress of the Province" by favoring one class of people – the European speculator – and one commodity – tea – over everything else. Richard Drayton has persuasively argued in this context that the "secular utopia" of the Improving agenda depended crucially on the market, and had "at its heart the theory that Nature was best used to yield commodities which might be traded widely, rather than to support local subsistence."[44] In our case, Jenkins and Mills used the alibi of tea's "special" status to lay out an elaborate land revenue and taxation scheme to "reform" and "modernize" Assam's agrarian economy. In essence, however, this "god of progress" raised the minimum land grant ceiling; increased the excise duty on opium – a drug allegedly at the root of the Assamese peasant's "indolence" and reluctance to work[45], including on these plantations; hiked revenue rates on all peasant lands and crops except tea; and doled out large holdings to tea planters without regard to their fiscal health or capacity to cultivate. Consider, for instance, that though the total government land revenue demand increased from 1 million rupees in 1865–66 to more than 4 million rupees in 1897–98, the growth of cultivated acreage under all varieties of crops *except* tea in Assam remained as low as 7 percent.[46] Similarly, for the 642,000 acres of land settled with planters between 1839 and 1901, over 85 percent was on concessional or privileged terms. For 1893, out of the 595,842 acres held by the European planters, 55 percent was under the revenue-free scheme and the other 30 percent at concessional rates far lower than what Assamese peasants paid for lands of similar quality.[47] Not all of Jenkins's original plans took root in these plantations, however. As indicated above, his idea of cajoling, or coercing local Assamese peasants into these plantations by raising taxes never really succeeded.

---

[43] Quoted in Mills's *Report*, appendix C, page xiv.
[44] See Richard Drayton, *Nature's Government: Science, Imperial Britain, and the "Improvement" of the World* (New Haven, CT: Yale University Press, 2000), p. 87.
[45] On the opium question, see Amalendu Guha, "Imperialism of opium in Assam 1773–1921," *Calcutta Historical Journal*, Vol. 1, No. 2 (January–June 1977): 226–245 and Shrutidev Goswami, "The Opium Evil in Nineteenth Century Assam," *Indian Economic and Social History Review*, Vol. XIX, No. 3 & 4 (1982): 365–376.
[46] Cited in the note by Chief Commissioner J. H. Cotton to the Government of India dated 1898, reproduced in *The Colonization of Wastelands in Assam*, and reprinted in Amalendu Guha, "Assamese Agrarian Relations in the Later Nineteenth Century: Roots, Structure and Trends," *The Indian Economic and Social History Review*, Vol. XVII. No. 1 (January–March 1980), p. 51.
[47] Guha, "Assamese Agrarian Society", p. 53.

If "improvement" was tea's gift to Assam, it came at a steep price for her local inhabitants and workers. As it is, not everyone was invited to partake of its promised munificence. Jenkins and Mills's land colonization scheme ensured that Assamese stakeholders in tea manufacture, with their limited capital and purchase money, were effectively kept out for a large part of the period under study. For the most part, local participation in the tea enterprise remained at the level of accountants, overseers, clerical employees or petty suppliers to European-controlled tea gardens and agency houses. While these circumstances made for a "Planter's Raj" to a great extent, this book shows that that "empire" was not just a congeries of all-knowing agents, and left much more than socio-economic upheavals in its wake.

Imperial Disarray: A Methodological Critique

> If the means of grace are employed, may we not also hope that [Assam] will become a garden of the Lord?[48]

The political circumstances of Assam tea's "discovery" in the backdrop of the First Opium War, its long imperial pedigree, and its subsequent validation as a distinctive commodity of British India were hailed as nothing short of Biblical miracles. For the East India Company (EIC), colonial bio-prospectors, and scientific "experts," however, botanical serendipity soon transformed into a narrative of bio-imperial conquest, natural "mastery" and social ascendancy in an erstwhile "wild" and unruly land. To that end, tea was the catalyst and tea estate the metaphor of an improving, Edenic regime. As Carolyn Merchant argues, the "Recovery of Eden" is one of Western culture's oldest "myths" – a meta-narrative of turning "wilderness into garden, 'female' nature into civilized society, and indigenous folkways into modern culture."[49] Or, as the prominent Canterbury Anglican Henry Sewell put it in the early 1850s, "the first creation was a garden, and the nearer we get back to the garden state, the nearer we approach what may be called the true normal state of Nature."[50]

---

[48] This phrase was used by Nathan Brown, the American Baptist missionary on his travels to upper Assam in 1836, quoted in H. K. Barpujari, Ed. *The American Missionaries and North-East India, 1836–1900* (Guwahati: Spectrum, 1986), pp. 7–8.

[49] Carolyn Merchant, *Reinventing Eden: The Fate of Nature in Western Culture* (New York, NY and London: Routledge, 2003), p. 2; see also, John Prest, *The Garden of Eden: The Botanic Garden and the Recreation of Paradise* (New Haven, CT and London: Yale University Press, 1981).

[50] Cited in W. D. McIntyre, ed., *The Journal of Henry Sewell, vol. 1. February 1853 to May 1854* (Christchurch: Whitcoulls, 1980), p. 427.

The analytics of "disarray" or unkemptness in this book, however, goes beyond providing a counter-narrative to this Enlightenment dictum of biosocial order. Rather, it is used as a heuristic framework to highlight ideological, material, and discursive inconsistencies, consequences, and contradictions of this plantation form and its purported mandate of agrarian development in the region. Indeed, neither the determinism of "recovery" nor the predictability of "resistance" tells the full story of *this* Eden. Instead, I show how the relationship between these two aspects of bio-imperialism moved in an expedient, disorderly and often contradictory fashion. To that end, scientific "rationality," epidemiology, agronomic planning, human management, modernity, and labor resistance were all inter-dependent nodes of this commodity history. Long hobbled by a historiographic approach that kept the plant on the one side and the plantation on the other, this syncretic look at the tea story shows that pathogens induced (both plant and human), laws transgressed, ideologies jettisoned, forests felled, and labor impoverished were, in fact, closely connected in Assam. It allows us to examine not just *what* went wrong in this "Recovery" story, but *how* things were allowed to go wrong, intentionally or otherwise.

This eco-social methodology also reveals other assumptions of the tea "modernity" narrative. As a legate of Whiggish and physiocratic agrarian doctrine, tea carried over many of its ideological burdens and political vocabulary. Touting its "agricultural" character as expedient, the tea enterprise invoked the emancipatory connotations associated with such pursuits since the reign of George III and Pitt the Younger. In its colonial avatar, these plantations became part of England's moral and material "duty" and "right" to transform and better nature. As Drayton suggests, a central axiom of this version of British imperialism was the idea of "colonization as amelioration."[51] The purported modernity of agrarian proselytism in the colonies fell squarely within these notions of responsible authority and "a mission towards which government might legitimately expand its powers."[52] But individuals and institutions (Sir Joseph Banks and the Kew Gardens, for instance) provided only one set of bulwarks and limits for this modernist project. For instance, David Arnold has persuasively argued that the impact of this agrarian credo on peasant agriculture in colonial India was not entirely clear, if at all.[53]

---

[51] Drayton, *Nature's Government*, p. 92. [52] Ibid., p. 89.
[53] See David Arnold's critique of Drayton in "Agriculture and 'Improvement' in Early Colonial India: A Pre-History of Development," *Journal of Agrarian Change*, Vol. 5, No. 4 (October 2005), and Richard H. Grove, *Green Imperialism: Colonial Expansion, Tropical Island Edens and the Origins of Environmentalism, 1600–1860* (Cambridge: Cambridge University Press, 1995).

Using the example of the Agricultural and Horticultural Society of India (AHSI), Arnold questions the limitations of this union of imperialism and improvement in the subcontinent. Ostensibly set up to foster evangelical ideas of progress and agrarian innovation, Arnold shows that AHSI's role in horticultural development remained mostly at the level of a "depository of practical information"; it rarely translated into matters of policy transformation or as a major force of empirical innovation. He would thus conclude that "Improvement and imperialism did not operate, as Drayton's argument might lead us to suppose, entirely in tandem."[54] Similarly, historian Richard Grove observes that the "utilitarian" science of colonial expansion and tropical garden Edens often coexisted with paradoxical concerns for planetary degradation and need for conservation.[55] He thus questions "monolithic" ideas of ecological imperialism by looking at the "heterogeneous and ambivalent nature of the workings of the early colonial state."[56]

Rather than attributing Improvement's working logic to a few individuals or institutions, or even to a well-planned colonial administrative policy, this book argues that its promise was critically sustained by a largely unseen – and historiographically ignored – ideological maneuver. As far as tea in eastern India is concerned, this was the function of an elaborate artifice. It emerged from Britain's purported ability to *separate* nature from nurture, to hold off that which was unregulated from that which the hand of man "improved." In other words, maintaining this schism between ecological indeterminacy (the nonhuman) and economic order (the human) was essential for tea's "progressive" and modernist agenda. As Bruno Latour argues, the presumption of Western modernity is based on a partition between two sets of practices: the mixed networks of the natural and human worlds on the one hand, and its assumed critical ability to ontologically separate one from the other.[57] If this "double separation" allowed for a plethora of post-Enlightenment practices –

---

[54] Arnold, "Agriculture and 'Improvement'," p. 516.
[55] See Grove, *Green Imperialism*. He also asserts that "while encouraged by the state, ostensibly for economic and commercial reasons, the botanical garden continued to encompass less openly expressed notions of tropical environment as a paradise, botanical or otherwise, which most professional botanists were keen to protect," in *Green Imperialism*, p. 409.
[56] Ibid., pp. 2, 7–8.
[57] See Bruno Latour, *We Have Never Been Modern*, trans. Catherine Porter (Cambridge, MA: Harvard University Press, 1993), especially chapters 1 and 4. Also see Donna Haraway, "The Promises of Monsters: A Regenerative Politics for Inappropriate/d Others," in Lawrence Grossberg, Cary Nelson, and Paula Treichler, eds., *Cultural Studies* (New York, NY: Routledge, 1992), pp. 295–337, and Haraway, *Modest_Witness@Second_Millenium.FemaleMan©_Meets _OncoMouse*™ (London: Routledge, 1997).

colonialism, imperial takeovers, or dichotomies of the "pre-modern" and secular, noumena and phenomena – to operate, Latour suggests that that constitution of "modernity" fundamentally relied on, yet rejected, the many mediations and hybrid in-betweens of nature and society, human and nonhuman, object and interpretation.[58] I invoke Latour's critical stance not just to highlight these networks[59] but to argue that Improvement's working principle in plantation economies was based on a similar tactic of separation: to absorb, utilize, and oftentimes create the disorderly while foreclosing it in its advertised mandate of agrarian order.

As far as tea is concerned, the following chapters show that the enterprise's attempts at such a separation was wishful thinking at best and an administrative, ideological, and horticultural oxymoron at worst. For not only could tea not be produced according to predictable and controllable natural rules, its economic success was dependent on manipulating the human/nonhuman divide in expedient and illegal ways. That human "ordering" *created* natural imbalance (as explored in Chapters 3 and 4) provides an ever greater, and damning, indictment of ideas of agrarian modernization.

"Unruliness," therefore, does not refer to the state of nature that this plantation economy "domesticated" and transformed. Nor is it invoked to merely indicate the damaging ecological effects of this monoculture agro-business. If capitalism's origin story moves from desert wilderness to cultivated garden, the following chapters highlight the many unseen in-betweens and ideological inconsistencies of that modernist parable.[60] Here, disorder is an integral functional component of how this plantation enterprise was run (or supposedly needed to make it run) – it does not pertain only to the condition of the landscape that tea took over, or left behind. In that vein, I question a central axiom in the historiography of "Improvement": that there was a state of natural "wilderness" that this Enlightenment order set out to reform in the colonies. In other words, this condition of indigenous disorder has been always read as an *object* for this imperial doctrine to act upon, successfully or otherwise – never as a *component* of Improvement's self-avowed mission.[61]

---

[58] See Latour, *We Have Never Been Modern*, especially chapter 2.
[59] To be sure, I do not subscribe to all of Latour's heuristic frameworks; for one, the question of "relativism" in his analyses of hybridity is problematic, as is the lack of analytic space for responsibility and accountability in social actions; see Chapter 3 for my engagement with critiques of Actor–Network Theory (ANT).
[60] See Carolyn Merchant, *Reinventing Eden*, p. 59; also Latour, *We Have Never Been Modern*.
[61] See, for instance, Drayton's position in *Nature's Government*; also see Alfred Crosby's *Ecological Imperialism: The Biological Expansion of Europe, 900–1900* (Cambridge: Cambridge University Press, 1986).

This book argues, instead, that disorder was an ally of Improvement, and a necessary one. It was non-regulation (or lawlessness) that allowed Edenic projects – in this case plant capitalism – to succeed. That it left behind messy legacies is a related matter. For instance, workers were transplanted and uprooted to Assam across great distances in the service of this cash-crop economy, often in contravention of legal stipulations.[62] Crowded and insanitary living conditions made for staggering labor mortality. Tea varietals were grafted from China to make the indigenous species more "refined" – only to result in disastrous market consequences. The industry's self-sanctioned role as harbinger of industrial modernity in eastern India was abandoned with expedient impunity to evade taxes. Laws were invoked and trampled to facilitate labor recruitment, manipulate wages, fell forests, acquire land, and keep out government intervention in tea matters. Monoculture tea ecology spawned legions of tea pests and bugs that, in turn, introduced "expert" entomological science and pesticide use – only to kill plant and people in its wake.

In terms of its analytic function, "disarray" in this book refers neither to a denigrated biological state nor a passive ecological metaphor. It points to the economic contrivances, the bureaucratic stratagems, the legal elasticity, and the agronomic manipulations that keep this pioneer plantation profitable and in the running. It is to this "productive" and continually operative state that unruly signals, to which the seemingly oxymoronic conjunction of the disorderly and the Edenic refers.

This methodology also adds to the debate on imperial history more generally.[63] Did empire, as construct and praxis, flow outward from Europe to her possessions, or was it a dialogic exchange of ideas, actors, and consequences? Dealing with the British Empire more specifically, scholars have deliberated whether it was best understood from the "vantage point of center or periphery"[64] or if these categories themselves needed rethinking.[65] Was empire a *product* of global history rather than its "driver?"[66] The move towards "new imperial history" signaled a desire to

---

[62] As a planters' representative in the Central Legislative Council of India noted in 1901: "when the coolie goes to a garden he begins to receive a wage and begins to live, whereas in Chotanagpur he only exists," quoted in Behal and Mohapatra, "Tea and Money Versus Human Life," Introduction.

[63] See the helpful overview, and references, on the shifts and "turns" in "old" and "new" imperial histories by Durba Ghosh, "Another Set of Imperial Turns?" in *American Historical Review*, Vol. 117, No. 3 (June 2012): 772–793.

[64] See William Beinart and Lotte Hughes, *Environment and Empire*, The Oxford History of the British Empire, Companion Series (Oxford and New York, NY: Oxford University Press, 2007).

[65] See Ghosh, "Another Set of Imperial Turns?" for an assessment of this historiographical shift.

[66] Ibid., p. 782.

Introduction 23

examine empires less through economic structures such as trade, commerce, and taxation but by its overall "culture" – through representations, discourses, language, and minority, feminist, and subaltern perspectives?[67] While the influence of poststructuralist thought and postcolonial theory on this "new" imperial history is unmistakable, a growing critique of this method highlighted the centrality of empiricism and the need to "go back to the archives." If the contemporary scholarly literature on the British Empire is any indication, empirical research – both as a method employed, and as a point of critique – provide, as historian Durba Ghosh terms it, another set of "turns" in the long historiography of empire.[68]

Whether or not empires moved centrifugally or centripetally, powerful forces combined towards the middle half of the nineteenth century to make the British Empire a lethal *bio-political* force.[69] The advances in transportation and communications, technological innovations and rapid industrialization, medical and military advances, and, in turn, a rising public demand of consumables and products that Britain neither produced nor possessed made for a global and spatial rearrangement of peoples, crops, floras, and landscapes. If the scholarship on imperial plantations and factories, and the transformation of the tropics more widely address this last issue,[70] this book contributes to the debate in two ways. It shows, firstly, that this ecological transformation was not a stable, end product of colonial power. Nor is it to be located in now-elapsed and archived ruins of colonial contact and resistance. I argue that the impacts of commodity capitalism launched during the apogee of British imperial power *continue* to be visible within the lives and

---

[67] See Antoinette Burton, "Thinking beyond the Boundaries: Empire, Feminism and the Domains of History," *Social History*, Vol. 26, No. 1 (2001): 60–71; Mrinalini Sinha, "Historia Nervosa or Who's Afraid of Colonial-Discourse Analysis," *Journal of Victorian Culture*, Vol. 2, 1 (1997): 113–122; Kathleen Wilson, "Old Imperialisms and New Imperial Histories: Rethinking the History of the Present," *Radical History Review* 95 (2006): 211–234; Catherine Hall and Sonya O Rose, eds., *At Home With the Empire: Metropolitan Culture and the Imperial World* (Cambridge: Cambridge University Press, 2006); Catherine Hall, ed., *Cultures of Empire: Colonizers in Britain and the Empire in the Nineteenth and Twentieth Centuries – A Reader* (Manchester: Manchester University Press, 2000); Kathleen Wilson, *A New Imperial History: Culture, Identity and Modernity in Britain and the Empire, 1660–1840* (Cambridge: Cambridge University Press, 2004). Also see Clare Anderson, *Subaltern Lives: Biographies of Colonialism in the Indian Ocean World, 1790–1920* (Cambridge: Cambridge University Press, 2012).
[68] See Ghosh, "Another Set of Imperial Turns?" especially pp. 788–793.
[69] See Corey Ross, *Ecology and Power in the Age of Empire: Europe and the Transformation of the Tropical World* for a recent assessment. Emphasis mine.
[70] On the idea of the tropics in ecological imperialism, see David Arnold, *The Tropics and the Traveling Gaze: India, Landscape, and Science, 1800–1856* (Seattle, WA: The University of Washington Press, 2006).

landscapes they impacted. In this sense, this book is part of an imperial historiography that sees empire's afterlives as a "protracted" and *ongoing* process of "ruination";[71] it seeks in this history of tea – primarily during its colonial, but also in its postcolonial phases – an active process of agro-ecological, social, and economic alteration and residues in Assam's lives and lands. The contemporary struggle with tea pests, persistent worker unrest and mortality, or even the emergence of ethnic disquiet and armed insurgency in the region are stark eco-social reminders of how imperial formations reside and transmit over a *long durée*.[72] This book inquires "how [...] imperial formations persist in their material debris, in ruined landscapes and through the social ruination of people's lives?"[73]

Finally, though this tea story forms part of the global environmental history of European agro-economic imperial exploitation, I am wary of looking at it through the predictable vocabularies of that historical field – namely, spatial change, environmental causation, degradation and conservation, or the relationship between metropolitan and indigenous knowledge systems.[74] While its chapters invoke some of these concerns, this book shows that, as far as imperial south Asia (and more importantly its easternmost "frontier") was concerned, there was no inherent predictability, legibility, or even legality to how eco-social plans of commodity production were formulated and operationalized. If the Whig credo of agrarian Improvement provided a crucial ideological referent as already remarked, tea making was largely a function of how law, landscapes, microbes, forests, and labor were expediently and haphazardly manipulated. Given the long existence of tillage and transhumance, settlement and mobility, ecological diversity and political power structures in the subcontinent, biopolitical conquest (as with other aspects of colonization) made for an uneven imperial experience for the British in India. As Mahesh Rangarajan observes, "the ecologies of forest, delta, and hillside were complex and unpredictable enough to confound foresters, hydraulic engineers, and civil officials ... the resulting tapestry ... is not easily amenable to homogeneity, whether imperial or

---

[71] See the recent work by Ann L. Stoler on this perspective in *Duress: Imperial Durabilities in Our Times* (Durham, NC: Duke University Press, 2016); also Stoler, ed., *Imperial Debris: On Ruins and Ruination* (Durham, NC: Duke University Press, 2013).

[72] Stoler, *Imperial Debris*, p. 5; see the Conclusion for a discussion of tea's contemporary legacies in Assam. Looked at this way, Jayeeta Sharma's *Empire's Garden* also contribute to this scholarship on imperial ruination, though her book is not theorized as such.

[73] See Stoler, *Imperial Debris*, p. 10. Also see Catherine Lutz's call for social ethnographies of *contemporary* empires in "Empire Is in the Details," *American Ethnologist*, Vol. 33, No. 4 (November 2006): 593–611.

[74] See, for instance, the categories used to elaborate the plan of Beinart and Hughes's, *Environment and Empire*, Introduction, pp. 1–21.

nationalist."[75] It is in this context that this book centrally argues for non-regulation as an integral component of tea's ecosystem and production ethos. It was this aspect of commodity imperialism in eastern India that helped it tide over little-understood plant ecologies, manage local habitats, ask for tax relief, negotiate conflicts over authority with the colonial government and area foresters, escape legal regimen, and condition pathogens environments and human lives with a constant eye towards profits.

Law, Disease, Disorder

Consider the function of law in these plantations. Despite being intimately linked to overall estate operations, legality was not just an institutional structure, a "symbol" of political authority, or vehicle to negotiate and manage the dilemmas posed by "White violence."[76] Legal stipulations in Assam exceeded these instrumental or ornamental functions. Here, law brought the colonial state and the tea enterprise into an uneasy, Janus-faced, and often paradoxical relationship. While the Calcutta administration used law to *facilitate* and pander planter requests for land acquisition, stricter labor control, forest use, and tax exemption, managers, on their part, often *blamed* colonial legislations and its regulatory parameters for unrealistic labor welfare goals, authoritarian overreach, and for creating economic and ecological roadblocks in the path of commerce. This expedient reliance on, and circumvention of legal provisions formed an integral part of the unregulated in the Assam plantations.

Law's fraught relationship between the adjudicator and the ruled is of course not new. From the EIC's early days, penal law attempted to reshape "exclusive sovereign rights" over the subject by appropriating, or in many cases extinguishing previously-held norms of rule, rank, status and gender.[77] If the Company espoused a more centralized mechanism to administer and codify its authority over people, these colonial ambitions were not wholly bereft of its reliance on symbols and substance of "traditional" authority. Indeed, the draft penal code (IPC) of 1837 was Thomas Babington Macaulay's attempt to shake

---

[75] See Mahesh Rangarajan, "Environmental Histories of India: *Of States, Landscapes, and Ecologies*," in Edmund Burke III and Kenneth Pomeranz, eds., *The Environment and World History* (Berkeley, CA: University of California Press, 2009), pp. 229–254.

[76] See Elizabeth Kolsky, *Colonial Justice in British India: White Violence and the Rule of Law* (Cambridge and New York, NY: Cambridge University Press, 2010) for an exposition of this last point of view.

[77] See Radhika Singha, *A Despotism of Law: Crime and Justice in Early Colonial India* (New Delhi: Oxford University Press, 1998), p. viii.

off these vestiges and paradoxes of "enlightened despotism." Though not operational till 1862, this new penal code aimed at breaking from the past – not a "digest of any existing system," but one based solely on the "universal principles of jurisprudence."[78] But as Radhika Singha argues, such ideals of standardization never really took off in colonial India. Law-making and law-giving remained a constant process of reworking idioms of power and "legitimate" political authority, and the methods through which they were to be communicated with the governed.[79] The many legal infractions between the colonial government and the tea lobby in Assam underscore these conflicts in matters of regulation and rule of law.

It has also been suggested that Macaulay's codification of Indian law was neither the result of an abstract English political philosophy nor solely aimed at instituting a "rule of State" but rather a response to – and an attempt to manage – the unruly "third face of colonialism," namely "non-official" White violence.[80] In this argument, law's function was both to "normalize" colonial brutality by according special rights and privileges to European British subjects, and simultaneously create a discursive space where these extra-legal exercises of power were challenged and, to some degree, accounted for. Read this way, the labor laws in the Assam plantations – invoked to ostensibly manage a "remote and irascible frontier" – effectively embolden, and effaced the use of managerial (including sexual) violence over the "coolie" labor employed; a protection *from* law and not *under* it.[81]

These legislations were indeed draconian, as labor historians have long argued. But what is underplayed is that between the government and the tea industry, the function of law was not necessarily an enabling one, nor did it always operate in each other's favor. In fact, for planters, the boundary between legal convenience and juridical overreach could be decidedly thin, especially when it came to matters of labor health and wellness. As Chapter 4 shows, these laws not only pandered to managerial demands for stricter labor control but also *regulated* norms of worker hygiene, well-being, sanitation and diet that planters were expected to follow. But in doing so, law highlighted the incommensurable, indeed paradoxical relationship between bodily fitness and profit margins that did not comfortably sit together. While disease prevention, worker housing and nutrition, and sanitary investments came

---

[78] Cited by the Indian Law Commissioners to the Governor-General in Council, October 14, 1837 in *Parliamentary Papers 1837–38*, Vol. 41, pp. 465–466, quoted by Singha, *A Despotism of Law*, p. 298.
[79] Singha, *A Despotism of Law*, Preface.   [80] Kolsky, *Colonial Justice in British India*, p. 8.
[81] Ibid., p. 158.

towards the bottom of planter priorities, the mounting labor mortality rates in Assam throughout the period under study indicate that, in this as in other respects, law was followed more in the breach than in practice. On their part, the industry often blamed legal over-ambition for *creating* unrealistic (read unprofitable) goals and parameters in terms of worker well-being that, in turn, wrecked an otherwise "healthy" state of affairs. As necessary, the government refuted, sided with, or looked askance at these warped medico-legal logics. For all its claims, law was never a stable or predictable referent of authority in the Assam gardens.

Deliberate, and oftentimes expedient mismanagement was a hallmark of the Assam plantations labor world. But as pointed out earlier, this book departs from the standard narratives of worker exploitation and capital-labor antagonism in these estates. For worker life was impacted and manipulated in many more way than direct physical control. In the case of labor health, for instance, mortality was not just a corollary of a visceral, overbearing planter regime;[82] it was implicated in, and resulted from *how* law, epidemiology, vector environments, and profiteering were negotiated in these estates. Similarly, when it came to statutory worker wages, planters often "read" legal wisdom with ingenuity and invidiousness to sidestep "guaranteed" minimum pay. That such unseen and unregulated extra-legal policies led to persistent worker impoverishment once again shows that law – or lawlessness – was a participant tool in the unkempt state of the Assam plantations; legality was invoked and trespassed as deemed profitable. In fact, depending on the circumstance, law could be expediently re-engineered or refashioned; trounced or trespassed; and celebrated or vilified by plantation functionaries, medical men, tea "experts," foresters, administrators and labor commissioners. In terms of its purview and reach, therefore, law in the Assam gardens served both a *maximal* function (as a structure of governance) and a *minimal* one (as a convenient tool of trade). In most times, these two purposes worked in tandem. But not infrequently, they were at odds with each other, as we shall see.

The discursive, ideological and instrumental role of law in the Assam gardens thus went much further – and was much more invidious – than simply aiding partisan agendas, whether official or non-governmental. In this case, juridical functions, and its extra-legal manifestations, extended to managing landscapes and labor, health and agronomy,

---

[82] See Rana P. Behal and Prabhu P. Mohapatra, "Tea and Money Versus Human Life: The Rise and Fall of the Indenture System in the Assam Tea Plantations 1840–1908."

state authority and planter control, litigations, and even literary dramatizations.[83]

Eco-social disorder was of course not limited to human manipulation alone. As with Jenkins's recommendations above, wastelands were continually doled out to bioprospectors at throwaway or nominal prices – often with regard to profitability or agrarian merit. Among others, these discriminatory land-grab and revenue policies had a cascading impact on private land ownership, immigration numbers, uncultivated but locked-up holdings, and disease environments in the province. Property relations between agriculturists and land in Assam were similarly modified with the government's new silvicultural agenda in 1864. The historical irony here is worth noting. As it turned out, the newly formed forestry department's mandate of husbanding and owning valuable woodlands was short-changed in northeastern India – for Jenkins's agrarian largesse had already parceled out much of these lands to tea companies earlier. The sometimes acrimonious, sometimes co-dependent, and many-a-times illegal dealings between foresters and planters fall squarely within this plantation's unregulated history.

Two primary tea pathogens – the *Helopeltis theivora* (or the tea mosquito bug) and the *Tetranychus bioculatus* (or the tea mite) and three human killers – malaria, cholera and kala-azar (or black-fever) complete the dramatis personae of these disheveled Edens. If human epidemiology has been occasionally examined in existing scholarship, tea pest bionomics has been largely outside the purview of social scientists and historians. This is both surprising and unhelpful. For disease environments, whether of humans or plants, were not visitations from the outside. They were intimately connected, and emerged out of the ecological, legal, and political economic structures of the tea enterprise. As it is, the persistent tea bug problem in these estates highlighted the limits to botanical "mastery" that underpinned the ideological foundations of this enterprise since the 1840s. Of course, pests also reminded planters that quantity was not an automatic

---

[83] Consider, for instance, that among other factors, Dakshinacharan Chattopadhyay's scathing critique of planter violence and harsh plantation life in Assam in his Bengali play *Cha-Kar Darpan Natak* in 1874 led to the promulgation of The Dramatic Performances Act the following year. See Pramila Pandhe, ed., *Suppression of Drama in Nineteenth Century India* (Calcutta: India Book Exchange, 1978); Tanika Sarkar, *Rebels, Wives, Saints: Designing Selves and Nations in Colonial Times* (New Delhi: Permanent Black, 2009); and Arnab Dey, *Of Planters, Ecology, and Labor: Plantation Worlds, Human History and Nonhuman Actors in Eastern India (Assam), 1840–1910*, unpublished PhD dissertation (Chicago: University of Chicago, 2012), chapter 1 for a discussion of this play and its historical contexts.

guarantor of profits – in diminishing taste and output, they remained a formidable and serious ecological competitor to the tea industry, and its constant boasts of techno-scientific "expertise" in eastern India.

Tea Culture: "Agrarian Imaginaries" and the Question of Sources

Culture, *n.*

I. the cultivation of land and derived senses;

2a. the cultivating or rearing of a plant or crop;

7b. a way of life or social environment characterized by or associated with the specified quality or thing; a group of people subscribing or belonging to this.[84]

This study of the Assam plantations has some interesting parallels with geographer Julie Guthman's critique of "agrarian imaginaries." Tracing the origins of this imaginary to the writings of Thomas Jefferson among others, Guthman argues that its "pastoral" vision eulogized the relationship between the cultivator and land as an essentially redemptive one. She suggests that such "righteous" narratives are regularly, if uncritically applied to alternative agricultural practices in the United States of America. Speaking for organic farming and consumption habits especially, Guthman contends that these ideas of "wholesomeness" absolves, if not obscures, its heavy reliance on industrialized, marginalized, and racialized labor (especially from the Latin American countries).[85] More recently, a study of the Darjeeling tea plantations of northeastern India recalibrates Guthman's phrase to argue for a "Third World agrarian imaginary" – an ascription of similarly uncritical ideas in the post-colony.[86] In this "third-world" guise, proponents of fair-trade, India's Tea Board officials, Darjeeling's Geographical Indication (or GI) champions, and advocates of the region's ethnically driven self-determination campaign (the Gorkhaland Movement) all invoke "return to a fictionalized distant past from which plantation societies have (unjustly) diverged."[87] This imaginary repositions the plantations as a "site" of these already-endured damages that the imputed "benevolence" and "environmental stewardship"

---

[84] *Oxford English Dictionary*, Third Edition, June 2008.
[85] See Julie Guthman, *Agrarian Dreams: The Paradox of Organic Farming in California* (Berkeley, CA: University of California Press, 2004).
[86] See Sarah Besky, *The Darjeeling Distinction: Labor and Justice on Fair-Trade Tea Plantations in India* (Berkeley, CA and London: University of Berkeley Press, 2014).
[87] Ibid., p. 30.

of planters, Gorkhaland activists, GI discourse, and fair-trade promises to amend and make good. This work contends that these tendencies both obscure and occasion the many monetary and non-monetary injustices that labor (especially female labor) encountered, and continue to encounter in their daily tasks of tea-making and manufacture. It suggests that the tolerable and intolerable aspects of the Darjeeling plantation world (and questions of justice and injustice) were forged not just through the "moral economic" antagonism between labor and capital, but also through a pragmatic reciprocity between workers, the management and its agro-environment.[88] The study bases its ethnography of this "tripartite" moral economy to examine the local strategies, relationships, visions, and choices that workers themselves drew up vis-à-vis the plantations.[89]

If there is ample room for a similar critique of Assam (especially organic Assam tea) – and this book does not do that – two caveats against the conceptual novelty of "Third World agrarian imaginary" and "tripartite moral economy" needs to be raised. For the first, an uncritical subscription to fair-trade's redemptive narrative may pose the problem of historical anachronism. While the rise of organic farming – whether in North America or under GI specifications in Assam or Darjeeling – may have re-energized these pastoral visions, they are by no means a symptom of the present. As this book shows, an expedient return-to-nature vocabulary, or paeans of planter-as-farmer, or the plantation-as-family discourse are scattered throughout the *colonial* history of tea making in northeastern India. Indeed, as Richard Grove has persuasively argued, these romanticized relationships between cultivator and land were always already present in the commercial exploitation of tropical garden Edens in the Old World and the New.[90] In the case of Assam tea, these "imaginaries" pre-date the coming of organic agro-marketing by many hundred years. I therefore use the word "culture" instead of imaginaries to discuss these ideas, and do so in a dual sense. Culture is used to mean (as in 1 and 2a above) both the agrarian practices of tea cultivation in the province, and (as in 7b above) to refer to the institutional character and political economic form of this enterprise. If the chapters below show the unregulated and disorderly relationship between this culture's seemingly extrinsic elements (such as bugs, law, taxation, nationalist politics) and its intrinsic components (land tenure, labor policies, and tea agronomics),

---

[88] The classic work on this topic is James C. Scott, *The Moral Economy of the Peasant: Rebellion and Resistance in Southeast Asia* (New Haven, CT: Yale University Press, 1976).
[89] See Besky, *The Darjeeling Distinction*, chapter 2, pp. 59–87.
[90] See Richard H. Grove, *Green Imperialism: Colonial Expansion, Tropical Island Edens and the Origins of Environmentalism, 1600–1860*.

the overall historical and historiographical burden of this book is to argue that these two aspects of the enterprise need to be read together.

Secondly, this book questions the heuristic efficacy of "tripartite" moral economy in clarifying connections between labor, management, and agro-environments. As with the previous issue, these relationships are often embedded within and in-between material, social, and ecological contexts that are difficult to separate. Of course, that economy was itself a product of ecology and terroir form part of this book's assessment of labor life in the colonial Assam gardens. To that end, the analytic use of a third (or fourth) determinant of the tea worker's "moral economy" may, once again, reproduce the binary of the plant on the one side and the plantation on the other, of colonial agrarian practices on the one hand and postcolonial policies on the other.

Finally, we come to the question of representing labor's *own* views about life on these estates. By extension, of course, this raises the issue of sources and this book's engagement with the problem. Besky's above work on Darjeeling claims that the "tripartite" moral economic framework helps us grasp how workers "themselves understand the plantation as a social and ecological form."[91] But how may one do this when, for all purposes, labor life in the Assam gardens were either effaced or always *represented for* in governmental and non-governmental reports, correspondences and memoirs? To rephrase a classic formulation – how may one "rescue" the "poor stockinger," the "Luddite cropper" or the "obsolete hand-loom weaver" in our colonial history when they were either not asked, or deigned too immature and "childlike" to be asked, in body and in intelligence, to be present in their own making?[92] While I have tried to "recover" their presence and points-of-view by reading in-between and against the official and non-official narrative grain,[93] this work avoids ventriloquizing, ascribing to contemporary worker testimonies a measure or reflection of former "truths," or reading labor intentionality back into these records.[94] For instance, the treatment of massive worker protests in

---

[91] Ibid., p. 3.
[92] See E. P. Thompson, *The Making of the English Working Class* (New York, NY: Vintage, rpt. 1966), Preface; I provide details of this conundrum in Chapters 3 and 5. Also see the critique of Thompson's work, especially on this question, by Carolyn Steedman, *Master and Servant: Love and Labour in the English Industrial Age* (Cambridge: Cambridge University Press, 2007).
[93] See Ranajit Guha's early call to imbibe this method in "The Prose of Counter-Insurgency," in Ranajit Guha and Gayatri Chakravorty Spivak, eds., *Selected Subaltern Studies* (New Delhi: Oxford University Press, 1988), pp. 45–88.
[94] See Gayatri Spivak's early doubts about "recovering" subaltern consciousness and the role of the archives in positioning subjectivities in "The Rani of Sirmur: An Essay in Reading the Archives," *History and Theory* 24.3 (1985): 247–272. Her classic intervention on this matter is "Can the Subaltern Speak?," in Cary Nelson and

these plantations during autumn 1920–21 in Chapter 6 is interested not so much in what they *felt* or *thought* during these events – for we have very little access to their state of mind or undistilled voice – but in the managerial nexus of agronomic, legal, and wage manipulation that the walkouts revealed. Indeed, labor pushback in the Assam estates, whether in the form of protests, desertions, absenteeism, or outright riots, highlights much more than an epiphenomena of their progress towards proletarianization.

To that degree, this book's reliance on what may be termed "elite" sources is not a selective use of materials. Similarly, if local peasants and plantation functionaries, and "commoners" associated with the industry appear infrequently in these pages, it is not because they are unimportant or extraneous to our story. While much of the story of labor subalterneity or of local, vernacular response to the Assam plantations have already been told,[95] this book's aim has been to unearth new and unexplored connections between land, labor, and landscapes; tea and terroir; ideological presuppositions and material impacts; and between tea's economic, social, legal, and environmental entanglements in northeastern India. As it is, readers will note that for the greater part of the nineteenth and the early twentieth centuries, local participation in everyday plantation operations was severely restricted, if not co-opted within the enterprise's commercial ambitions. Ultimately, if tea – and its purported "improving" regime – was a "God that failed"[96] local Assamese aspirations and sensibilities, this book reminds readers that it also left a troublesome legacy on the life and limbs of the millions involved, and on the trees and terrain of the land that spawned its growth and success. This work highlights the colonial origins of these problems and their continuities into the present.

Chapter 1 provides an overview of tea's long imperial pedigree leading up to its purported "discovery" in British east India during the First Opium War. The ideology of botanical, scientific triumphalism and progress surrounding this "find" – and its many contradictions – is examined in Chapter 2 by looking at a long-drawn tea income-tax debate between the Government of India and the planting community during the late-nineteenth and early twentieth centuries. These macroeconomic issues are substituted for more pressing, yet invisible micro-ecological concerns in Chapter 3. Bringing the plant and the plantation together, this chapter discusses tea pests and their impact on production, labor life, profits, and

---

Lawrence Grossberg, eds., *Marxism and the Interpretation of Culture* (Urbana, IL: University of Illinois Press, 1988), pp. 271–313.

[95] See the discussion above on Amalendu Guha's *Planter Raj*, Jayeeta Sharma's *Empire's Garden*, and the numerous works on Assam tea labor in footnote. 8.

[96] This is Jayeeta Sharma's phrase; see *Empire's Garden*, p. 13, and part I.

the discourse of botanical "expertise." Chapter 4 focuses on three human diseases (namely cholera, malaria, and black fever) that caused staggering labor mortality in these estates. Unlike epidemiological explanations centered on therapeutics and public hygiene, this chapter argues that our assessment of labor health and morbidity in the Assam gardens need to take pathogenicity, profiteering, and legality simultaneously into account. If human management was important for this cash-crop enterprise, landscape manipulation was not too far behind. Chapter 5 discusses the relationship between the forest department and the tea enterprise to show how fiscal exigencies of these two major resource stakeholders occasioned extraordinary, unregulated, and extralegal tweaks to the paradigm of "rational" scientific silviculture in British east India. As we turn to the next century in Chapter 6, ecology and economy converge in making tea a highly political plant. In the backdrop of Gandhi's first noncooperation call, this chapter examines a series of worker protests that erupted in the Assam gardens around 1920–21. I shift the focus from causal explanations of this event in existing accounts to examine *how* a peculiar "culture of commerce" – centered on law, crop agronomics, and wage tampering – converged in the making of worker impoverishment and anger in the run-up to these walkouts. The Conclusion shows how the agro-economic, social, and ecological costs and consequences of tea's colonial origins explored in this book remain visible to this day.

1 Planting Empires

A discovery has been made of no less importance than that the hand of Nature has planted the shrub within the bounds of the wide dominion of Great Britain: a discovery which must materially influence the destiny of nations; it must change the employment of a vast number of individuals; it must divert the tide of commerce, and awaken to agricultural industry the dormant energies of a mighty country.[1]

Tea is no ordinary plant. Appropriated by global empires, and dated to more than two thousand years in antiquity, *Camellia sinensis*'s (or tea's) fortunes were closely linked to the ebb and flow of personal ambitions, court intrigues, oceanic commerce, and imperial desires. It is the stuff of lore – oral, documented, imagined, and conjured. This chapter explores tea's early history leading up to its purported "discovery" in the easternmost corner of British India around the mid-nineteenth century.[2]

Samuel Ball notes that the earliest "legendary" allusion to tea is ascribed to the reign of a Chinese emperor, Shen Nung, around 2737 BCE in his *Pen ts'ao*.[3] Ball is quick to comment, however, that this was an anachronistic reference – the tract being composed only during the Neo-Han dynasty of AD 25–221, and the tea citation added after the seventh century when the word *ch'a* came into use.[4] William Ukers in his

---

[1] G. G. Sigmond, *Tea: Its Effects, Medicinal and Moral* (London: Longmans, 1839), p. 144.
[2] I draw on the following accounts of tea's early origins: William H. Ukers, *All About Tea*, Vol. I (New York, NY: The Tea and Coffee Trade Journal Company, 1935); Samuel Ball, *An Account of the Cultivation and Manufacture of Tea in China* (London: Longman, Brown, Green, and Longmans, 1848); Samuel Baildon, *Tea in Assam: A Pamphlet on the Origin, Culture, and Manufacture of Tea in Assam* (Calcutta: W. Newman & Co., 1877); Samuel Phillips Day, *Tea: Its Mystery and History* (London: Simpkin, Marshall & Co., 1877); David Crole, *Tea: A Text Book of Tea Planting and Manufacture* (London: Crosby Lockwood and Son, 1897); Sir Percival Griffiths, *The History of the Indian Tea Industry* (London: Weidenfeld and Nicolson, 1967); J. Ovington, *An Essay Upon the Nature and Qualities of Tea*, Second edition (London: Printed for John Chantry, 1705), and William T. Rowe, *China's Last Empire: The Great Qing* (Harvard, MA: Belknap Press of Harvard University Press, 2009), especially chapter 5.
[3] Ball, *An Account of the Cultivation and Manufacture of Tea in China*, p. 1.
[4] See Ukers, *All About Tea*, p. 2 for a discussion of this debate.

Figure 1.1 Tea plant (*Camellia sinensis*): flowering stem. Watercolour ©
Wellcome Library

encyclopedic work suggests that the "earliest credible mention" of tea is probably to be found in the fourth century under the Chin dynasty.[5] It has been argued that after the third century of the Christian era the plant's references become more numerous and "seemingly more reliable."[6] In any case, by the fifth century we begin to find tea being "traded" as an article of exchange. Ukers notes that in Chiang Tung's *The Family History of Chiang* of the Northern Sung dynasty (AD 420–79), the sale of vinegar, noodles, cabbage, and tea was taken as a reflection of the dignity of the government.[7] If tea's medicinal roots are well documented in these

[5] Ibid.   [6] Ibid., p. 3.   [7] Ibid.

accounts, its transition to a "refreshing beverage" is also alluded to in Chinese sources by the sixth century. By AD 780, an "exclusive work" was published by Lu Yu, a noted Chinese tea expert, who commented on tea's horticultural aspects, among others. By the time of the Sung dynasty (AD 960–1280), the drink had reportedly become widespread throughout all provinces in China.

Though these circumstances – and the evidentiary base – lack full historical credence, it is generally accepted that the "genesis" of tea cultivation, trade, and usage is firmly located in China. Its botanical origins, however, are more contested and controversial.[8] For instance, Samuel Baildon theorizes that the plant was introduced to China and Japan from India some 1,200 years ago.[9] Sir Percival Griffiths, in his compendium history of tea, however, argues that tea was indigenous to China, though he agrees with the Japanese side of this debate.[10] Despite these claims, a definitive pronouncement on the topic is well nigh impossible. Botanist C. P. Cohen Stuart argues that, contrary viewpoints notwithstanding, the answer to this age-old enigma is to be found amidst "Mother Nature" in the borderlands of China – the "mysterious Tibetan mountain walls, and the scarcely explored jungles of southern Yunan and Upper Indo-China."[11] An ecological explanation of tea's botanical roots is also sometimes proffered.[12] In this version, the monsoonal regions of South and Southeastern Asia, with its ideal soil, climate, topography, and rainfall, make for a "primeval tea garden" for the "natural propagation" of the plant – whether wild, native, or hybrid.

It is now widely accepted that tea's introduction to Europe did not happen till around 1610, when the Dutch brought the article to its shores. Textual references to the plant, though, make their first appearance around 1559 in Venetian author Giambattista Ramusio's *Navigatione et Viaggi*.[13] Known as the "Hakluyt of Venice," Ramusio was reportedly introduced to tea by the Persian traveler and merchant Hajji Mahommed (or Chaggi Memet). To be sure, the opening up of a speedy and efficient trade route through the Cape of Good Hope in 1497 brought European traders, pioneered at this time by the Portuguese, into closer contact with these eastern "cultures of tea." From Malacca in 1516 to Macao and thereafter to Japan in 1540, Portuguese traders, Catholic missionaries, and travelers carried the legend, lore, and lure of tea to Europe and

---

[8] Ibid.
[9] See Baildon, *Tea in Assam: A Pamphlet on the Origin, Culture, and Manufacture of Tea in Assam*, pp. 8–10; also Ukers, *All About Tea*, p. 6.
[10] Griffiths, *The History of the Indian Tea Industry*, p. 7.
[11] Ukers, *All About Tea*, pp. 6–7.   [12] See ibid. on this point.
[13] Griffiths, *The History of the Indian Tea Industry*, pp. 14–18; Ukers, *All About Tea*, p. 23.

beyond. The great Continental scramble for this prized beverage was about to begin. Father Gasper da Cruz, a Catholic priest, is credited with publishing the first Portuguese note on tea in 1560 upon his return from China.[14] Similarly, Father Louis Almeida published an Italian account of his experience with the drink in Japan in 1565.[15] Two years later, in 1567, Russian travelers Ivan Petroff and Boornash Yalysheff "casually noted" that the tea plant was a true wonder of China.[16] These narratives – whether extensive, effusive, or cursory – continue in Italian, Portuguese, French, and Dutch travel and missionary accounts throughout the sixteenth and seventeenth centuries. Though all aspects of this history need not detain us here, it is important to note that the salutary, medicinal, and invigorating aspects of the plant always find mention in these tracts. In other words, the wholesomeness of the product – for body and mind – is interspersed with details of its social, botanical, and commercial intercourse in the East. Father Alexander de Rhodes, in his *Voyages et Missions Apostoliques* (published in Paris in 1653), notes: "one of the things contributing to the great health of these peoples [the Chinese], who frequently reach extreme old age, is tay, which is commonly used throughout the Orient."[17]

Ecclesiastic voyagers aside, imperial trade rivalry contributed most to tea's circulation and exchange during this period. The Portuguese had an early and decisive advantage over commercial sea routes till about 1596. On their return with spoils from the East, Dutch ships carried goods from Lisbon to ports in France, the Netherlands, and the Baltic. Dutch navigator Jan Hugo van Linschooten noted with exasperation in 1596 that his country had not fully partaken in its share of this "rich oriental trade" item. Linschooten's account is credited as the first to take notice of tea (as *chaa*) in the Dutch language, and was translated into English in London in 1598.[18] As the Dutch consolidated their commercial moorings, especially with the founding of the Dutch East India Company (VOC) in 1602, the imperium saw the arrival of the British in this lucrative business of spices, among others. While details of the Anglo-Dutch rivalry around the Indian Ocean rim do not concern us here, its overlap with the tea story is interesting. Thus, the "massacre of Amboyna" in 1623 over territorial rights led to the English East India Company (EIC) – chartered in 1600 – acceding to Dutch claims over the spice island, and the simultaneous retreat to mainland India and adjoining areas. Despite arriving in English ships in compliance with the Navigation Act of 1651, therefore, the first

[14] Ibid., p. 24 [15] Ibid., p. 25. [16] Ibid. [17] Ibid., p. 27.
[18] Ibid.; also, Robert Parthesius, *Dutch Ships in Tropical Waters: The Development of the Dutch East India Company (VOC) shipping network in Asia, 1595–1660* (Amsterdam: Amsterdam University Press, 2010).

teas used in England around 1657 were all from Dutch sources. It is known that the first sizable importation of tea by the EIC was in 1669, when 143½ pounds of the article was sent to London from Bantam, Java.[19] Imperial rivalries around tea were to become much more serious within the next two centuries.

As it is, tea's early history in the British Isles was rocky and filled with opposing viewpoints. Though the first printed reference to the drink dates back to 1598 in *Linschooten's Travels*, an English translation of a work originally published in Holland, a Scottish physician, Dr. Thomas Short, argues that it may have been known as far back as the reign of James I, for the first East India fleet sailed around 1601.[20] While these accounts remain inconclusive and disputed, it is indeed remarkable that the EIC did not make a more forceful stake over the article's commerce vis-à-vis its Dutch competitors. By the time English diarist Samuel Pepys notes about sending for "a cup of tee,"[21] on September 25, 1660, the article had already acquired some familiarity, at least within the precincts of the English coffee houses and in the EIC fraternity. As the debate on tea's curative versus intoxicating effects continued, Act XXI of Charles II, c. 23 and 24 taxed the product for the first time with an excise duty of eight pence for every gallon of tea sold. In 1669, imports of tea from Holland were prohibited, thereby creating an early monopoly for the EIC to trade in the commodity. For tea writers, however, it was the arrival of the Portuguese princess, Catherine of Braganza, as Charles II's wife – and as Queen of England – that solidified the drink's fame and fortunes within British social circles. David Crole notes that from a mere two pounds of the "rare delicacy" presented to her by the EIC in 1664, the Company went on to import close to 4,713 pounds a century later.[22] Catherine, however, is often credited as England's "first tea drinking queen" and a patron of its "temperate" qualities in place of ales, wines, and other spirits.[23] As English social custom, coffee houses, and drawing rooms picked up this habit through the seventeenth and the eighteenth centuries, there was no looking back for tea's elevation as Britain's "national drink."[24]

---

[19] Quoted in Ukers, *All About Tea*, p. 29.   [20] Cited in Ukers, *All About Tea*, p. 37.
[21] See ibid., pp. 40–41 for details on Pepys; also see Sigmond, *Tea: Its Effects, Medicinal and Moral*.
[22] Crole, *Tea: A Text Book of Tea Planting and Manufacture*, p. 19.
[23] Ukers, *All About Tea*, p. 43.
[24] See Julie E. Fromer, *A Necessary Luxury: Tea in Victorian England* (Athens, OH: Ohio University Press, 2008); also see Erika Rappaport's recent work, *A Thirst for Empire: How Tea Shaped the Modern World* (Princeton, NJ and London: Princeton University Press, 2017).

Planting Empires 39

Figure 1.2 The Great Tea Race © The National Archives Image Library, United Kingdom

It is indeed ironic that among all these notes on tea usage and circulation, India finds little mention. In fact, two historical accounts of its appearance during the Mughal Empire (1526–1857) appear to contradict each other. On the one hand, Johan Albrecht von Mandelslo, the German adventurer who visited Persia and India during the seventeenth century, seemed to have discussed tea drinking during his visit to the country in 1640. Mandelslo is reported to have observed: "at our ordinary meetings, every day, we took only *The*, which is commonly used all over the Indies."[25] A similar statement is also attributed to John Ovington in the latter's *Voyage to Suratt*, published in 1689.[26] However, in his magisterial history of agrarian systems of Mughal India, Irfan Habib discounts the possibility of tea being widely used, then "just coming to be known, but ... not cultivated anywhere, not even in Assam, where it must have existed in a wild state."[27] Habib cites Shihabuddin Talish's *Fathiya-i'Ibriya*, a seventeenth-century chronicle of what is Assam today, to argue that "nothing like tea is described or referred to in it."[28] Sir Percival Griffiths provides another curious interlude to this enigma. Griffiths suggests that, though references to tea began to appear in EIC's court minutes less than thirty years after Mandelslo's visit, they were all about consignments that were either "from the Far East to England, either direct or via India." He argues that these minutes throw no light on its use and consumption habits within India.[29] Though details of this historical dispute need not detain us any further, it is safe to say that tea-drinking, if not its availability in elite circles, was severely limited during the Mughal period. Our best sources for common life during the Mughal period – whether in Arabic, Persian, or in European languages – make no reference to the article's wide circulation and production in India during this period.[30] Indeed, Habib clarifies that "the tea and coffee plantations of today lie largely outside the limits of the Mughal empire."[31]

By the middle of the eighteenth century, however, EIC's trade affairs, especially with China, were beginning to show signs of political strain.[32] Internally, of course, EIC was a house in disarray, with allegation of wanton corruption, favoritism, "tea smuggling" and financial irregularities.[33] Indeed, in 1772 it "begged" the British government to

---

[25] Ukers, *All About Tea*, p. 134   [26] Ibid.
[27] Irfan Habib, *The Agrarian System of Mughal India 1556–1707*, Third edition (New Delhi: Oxford University Press, rpt. 2014), p. 51.
[28] Ibid., fn. 94.   [29] Griffiths, *The History of the Indian Tea Industry*, p. 12.
[30] See the Bibliography appended to Habib's monograph, especially section 1J, pp. 488–494.
[31] Habib, *The Agrarian System of Mughal India*, p. 62.
[32] For an interesting history see William T. Rowe, *Hankow: Commerce and Society in a Chinese City, 1796–1889* (Stanford, CA: Stanford University Press, 1992), especially chapter 4.
[33] For a general history of the EIC, see John Keay, *The Honourable Company: A History of the English East India Company* (New York, NY: Macmillan, 1994); Emily Erikson, *Between Monopoly and Free Trade: The English East India Company, 1600–1757* (Princeton:

remit the contributions it owed, and lend it £1,000,000 to steady its trade balance.[34] Eager to have a greater say in Company affairs, and infuse economic accountability, the EIC monopoly was revoked in 1813 by the British parliament. Interestingly, EIC's trade monopoly to China – consisting principally of tea commerce – was allowed to continue for another two decades, being finally abolished in 1833. EIC's reputation notwithstanding, the government collected £77,000,000 from it between 1711 and 1810 on tea alone.[35] By 1831, however, the Chinese authorities in Canton had become increasingly indignant at the Company's affairs – especially its clandestine importation of opium – and imposed restrictions on vessels, men, and movement. An imperial edict in 1832 ordered all Chinese maritime provinces to build ramparts and actively prepare ships to fend off European ships appearing on its coasts. If belligerence transformed into the First Opium War (1839–42),[36] commerce had other worries to contend with. With an importation figure of 32,000,000 lbs. from Canton in 1843,[37] tea stood preeminent amongst the articles traded by the British with the Celestial Kingdom. Though reticent earlier, the exigencies of war, the closure of treaty ports, and fear of fiscal ruin now forced the Company to consider an alternative source for this prized commodity. They did not have to look far.

"The Most Brilliant Discovery": The Assam Gardens in History

Though the timing of these upheavals with China was fortuitous, the plant had already moved around in political circles in India for some time. Sir Joseph Banks, for instance, endorsed tea cultivation as a favorable agrarian venture for India in his memoirs of 1788.[38] Colonel Robert Kyd of the Bengal infantry, and the first superintendent of the Calcutta Botanical Gardens (CBG), reportedly planted them in his then-private gardens at Shibpur in around 1780. We are also told that Edward Gardner, honorary resident at the royal court at Kathmandu, Nepal noticed some shrubs similar to tea growing in the palace precincts and

---

Princeton University Press, 2016); and Tirthankar Roy, *The East India Company: The World's Most Powerful Corporation* (New Delhi: Allen Lane, 2012).

[34] Ukers, *All About Tea*, p. 70.   [35] Ibid., p. 74.

[36] For a recent history of the Opium Wars, see Mao Haijian, *The Qing Empire and the Opium War: The Collapse of the Heavenly Dynasty*, trans. Joseph Lawson, Peter Lavelle, and Craig Smith (Cambridge: Cambridge University Press, 2016); also see Arthur Waley, *The Opium War Through Chinese Eyes* (Stanford, CA: Stanford University Press, 1958), and Roy Moxham, *Tea: Addiction, Exploitation, and Empire* (New York, NY: Carroll & Graf, 2003).

[37] Ukers, *All About Tea*, p. 77.   [38] Ibid., p. 134.

sent them to Dr. Nathaniel Wallich, later director of the CBG, in 1816. These were returned as *Camellia drupifera* and not "true tea."[39]

Despite these false starts, military explorations beyond the easternmost limits of British India by 1823 inadvertently turned around tea's prospects in the region. Then forming part of Burmese territory – and inhabited by "hill tribes" from the Muttock, Singpho, and Khampti communities – these regions of upper Assam shared several ecological characteristics with their southeastern neighbor and Yunnan. It is here that Major Robert Bruce, on a trade expedition with the local Singphos in 1823, reportedly "found" tea trees similar to those in China growing in wild abandon.[40] On a return visit, these were handed over to the Major's brother Charles – then commanding a division of gunboats in the Sadiya region during the first Anglo-Burmese war of 1825–26.[41] Duly apprehensive of their quality and authenticity, Charles Bruce forwarded some of these seeds to David Scott, then Agent to the Governor-General of the North East Frontier. As the civil war with Burma raged on, the fate of these peripatetic seeds rested unknown – and "scientifically" invalidated – for the next decade or so.

Meanwhile, momentous political changes were afoot in this eastern province of British India. Having driven the Burmese out of Assam in 1826,[42] the EIC swiftly moved to reinstall Purandhar Singha, the last descendant of the region's long-ruling Ahom dynasty, as the puppet prince of upper Assam. At his end, however, the new Governor-General, William Cavendish Bentinck, had already set the ball rolling for a renewed look at the question of tea planting. He laid before his Council on January 24, 1834 an "elaborate scheme" for its introduction and a twelve-member Tea Committee was formed that included Nathaniel Wallich; James Pattle; a Chinese physician, Mr. Lumqua; G. J. Gordon; Raja Radhakanta Deb; and Ram Comul Sen, among others.[43] While Gordon set sail for China and Dutch Java on the *Water Witch* to gather information and specimens from

---

[39] Ibid., p. 135.
[40] House of Commons Parliamentary Papers, 63, No. 63, 1839; though all tea histories attest to Bruce's report, there is lack of clarity as to the exact nature of *his* sources; see Crole, *Tea: A Text Book of Tea Planting and Manufacture*; Ukers, *All About Tea*; H. H. Mann, *Early History of the Tea Industry of Northeast India* (Calcutta: General Printing Co. Ltd, 1918); H. K. Barpujari, *Assam: In the Days of the Company 1826–1858* (Gauhati: Lawyer's Book Stall, 1963); and Jayeeta Sharma, "British science, Chinese skill and Assam tea: Making empire's garden," *Indian Economic and Social History Review* 43 (2006): 429–455 for an assessment of this question.
[41] See "Papers Relating to the Burmese War," House of Commons Parliamentary Papers, No 6, February 1825.
[42] See Barpujari, *Assam: In the Days of the Company*, and "Treaty of Peace with His Majesty The King of Ava," House of Commons Parliamentary Papers, No. 1, February 24, 1826.
[43] Ukers, *All About Tea*, p. 138.

these tea-producing countries, reports of "native" tea plants growing in the regions previously visited by the Bruce brothers flooded Calcutta once again. Among these informants, Lieutenant Andrew Charlton of the Assam Light Infantry and officer-in-charge of Sadiya remains preeminent. In 1831, he sent in "three or four young plants" found growing "wild" in Beesa to the Calcutta botanical authorities. Charlton also observed, like Bruce, that the local Singphos and Khamptis were "in the habit of drinking an infusion of the leaves."[44] Though unlucky in making his case at the time (his consignment to Calcutta reportedly died out ere long), Charlton had come to the notice of Scott's successor, Captain (later General) Francis Jenkins.[45] Responding to a circular sent out by the Tea Committee on March 3, 1834 soliciting opinion regarding the plant's prospects in British India, Captain Jenkins – with help from Charlton – sent in a "complete exhibit, which included tea leaves, fruit, blossoms, and the prepared leaf used by the hill tribes for making their primitive tea drink."[46] Though the semantics of primitivism was important, as we shall soon see, Jenkins's strong recommendation that the shrub was truly a local, and not an imported, product struck a chord with the Calcutta authorities. Dr. Wallich reportedly examined it in November that year and expressed satisfaction with its botanical pedigree. The Tea Committee exulted to the government:

> It is with feelings of the highest satisfaction that we are enabled to announce to his Lordship in Council that the tea shrub is beyond all doubt indigenous in Upper Assam ... we have no doubt in declaring this discovery ... to be by far the most important and valuable that has ever been made in matters connected with the agricultural or commercial resources of this empire.[47]

The Committee also expressed confidence that "under proper management," the plant's success for commercial purposes was beyond doubt. Losing no time in the matter, the government dispatched a "scientific mission" to Assam in 1836 under Dr. Wallich's superintendence. The team included geologist J. McClelland and botanist Dr. William Griffith. At Sadiya they were joined by Charles Bruce, who acted as guide and informant. Between January 15 and March 9, 1836, five

---

[44] Ibid., p. 137.
[45] Colonel Jenkins was in a sense the first British official who proposed the large-scale initiation of a plantation economy in Assam using sugarcane, mustard, mulberry, and indigo, among others. It is recorded that Jenkins had proposed that "the first duty of the Government ... is to make monopoly impossible ... that the great national tea-trade in Assam (should be) open to all, as the indigo trade in Bengal," quoted in Barpujari, *Assam: In the Days of the Company*, p. 223.
[46] Ukers, *All About Tea*, p. 139.
[47] Quoted in Ukers, *All About Tea*, p. 139; also, Barpujari, *Assam: In the Days of the Company*, pp. 217–18.

localities were surveyed where the indigenous plant was found growing in profusion.[48] Dr. Griffith reported that the tea trees examined were remarkably vigorous, and were of all ages, from seedlings to mature. When seen in February 1836, most of the full-grown shrubs were covered with seed buds, and some still bore blossoms. The older leaves were found to be of a fine, dark green color.

Despite this early validation of the plant – and its future – in eastern India, the divided, indeed disorderly, opinion of the Tea Committee and government "experts" over the next three years show that, as far as tea was concerned, commerce and ideology did not always converge. As Londa Schiebinger argues: "Europe's naturalists not only collected the stuff of nature but lay their own peculiar grid of reason over nature so that nomenclatures and taxonomies ... often served as 'tools of empire'."[49] In our case, the question of the Assam plant's authenticity vis-à-vis its Chinese counterpart was not just a matter of botanical doxa alone; it had racial, civilizational overtones. Given the far-flung locale of its "discovery" and the purported "backwardness" of its indigenous first-users, the Assam plant underwent several rounds of "scientific," experimental vetting before qualifying as a worthy candidate of "refined" metropolitan palates. Indeed, Griffiths was unabashed in his opinion that "a wild plant is not likely to give as good produce as one that has been cultivated for centuries."[50] To this end, two issues had to be decided: the type of plant to be favored for cultivation – whether indigenous or Chinese – and the location of these experimental gardens in British India. "Expert" opinion of the matter was sharply divided, and had long-term consequences for the tea enterprise. Some, like Wallich, Dr. J. Forbes Royle, and Dr. Hugh Falconer (superintendent of the government Botanical Gardens at Saharanpur), recommended the northern Himalayan regions of Kumaon-Garhwal (in present-day Uttarakhand), whereas Griffiths and McClelland supported upper Assam as being well adapted for the plant's future prospects. Indeed, the first locality chosen in Assam for tea gardens – a riverine sandbank (or *chur*) at Koondilmukh – was a spectacular error of judgment. With shifting sands and shallow alluvial soil, these *churs* proved to be inadequate germinating ground for tea seedlings, which soon withered and died upon planting.[51] David Crole wryly remarked: "in a short time ... the kindly Brahmaputra flowed over the site of this ... and buried in its waters a lamentable failure."[52]

---

[48] Ukers, *All About Tea*, p. 140.
[49] See Londa Schiebinger, *Plants and Empire: Colonial Bioprospecting in the Atlantic World* (Cambridge, MA: Harvard University Press, 2004), p. 11.
[50] Ukers, *All About Tea*, p. 140. [51] Ibid., p. 141.
[52] Crole, *Tea: A Text Book of Tea Planting and Manufacture*, p. 26.

When it came to the first question, opinion was similarly divided. Griffiths and Wallich were disposed towards importing Chinese seeds for these experiments, but McClelland believed that the native plant could succeed on its own. As the battle over species and habitat continued throughout 1836, most members of the committee finally agreed that the "China plant and not the *degraded* Assam plant" be used.[53] If imperial self-sufficiency was an important consideration, adhering to racial typologies was an even greater concern. Mr. Gordon was again dispatched to China that year to collect tea seeds, and reportedly brought back enough to raise 42,000 shrubs at the CBG. Of these, around 20,000 were dispatched to the old hand Charles Bruce at Saikhowa near Sadiya, though only 8,000 survived the journey.[54] Bruce, who had succeeded Captain Charlton as the superintendent of tea culture in 1836, lacked "formal" botanical training, but his close knowledge of the tea country, acquaintance with "hill chiefs" and pioneering spirit were seen as ample compensation for more esoteric knowledge. In addition to the Singphos and Khamptis, Bruce was asked to liaison with Chinese growers in these experimental gardens who had been specially brought in for the purpose. From his headquarters near Jaypur the following December, Bruce sent twelve boxes of manufactured tea to Calcutta for onward transmission to England. The specimen was allegedly found to be "equally as good as that produced in China."[55] To bolster its chances, the government requisitioned the services of one Mr. Lum Qua, a noted Chinese interpreter and "tea expert," and added him to Bruce's establishment. With Lum Qua's help, other Chinese growers and laborers, especially from the Straits Settlement, Penang, and Singapore, were brought in to Assam, though their efficacy and racial "purity" as ethnically Chinese (and therefore suited for the purpose) were reportedly questioned by Griffiths.[56] By late 1837, however, Chinese involvement in the region's tea affairs – both in terms of men on the ground, and as the cross-pollinated or hybrid China-Assam plant – was beyond doubt. In 1838, ninety-five chests of black tea dispatched to London were enthusiastically vetted for their "strength, pungency and astringency."[57] Elated, the Tea Committee declared that Assam would soon be able to compete favorably with China for so "indispensable" an article of commerce.

While the Tea Committee's honeymoon with Chinese tea men and botanical hybrids was soon to prove short-lived, other political maneuvers

---

[53] Ukers, *All About Tea*, p. 140; emphasis added.   [54] Ibid., p. 142.
[55] Cited in Barpujari, *Assam: In the Days of the Company*, p. 220.
[56] See Jayeeta Sharma, "British science, Chinese skill and Assam tea: Making empire's garden," and Sharma, *Empire's Garden*, pp. 37–38 on this issue.
[57] Barpujari, *Assam: In the Days of the Company*, p. 220.

were underway to make way for these emerging gardens of Empire. Wanting to yield sole authority in this new commercial environment, the EIC moved to depose the last Ahom noble, Raja Purandhar Singha, from his throne in 1838 on charges of "misgovernment." Five years later, the Company quelled a rebellion by the Singphos at Sadiya, who had long resented British incursions into their lands in the name of tea.

If political cunning yielded some results, the botanical wisdom of Assam's early tea experts was not entirely propitious. The decision regarding the China-Assam hybrid is a case in point. For planters, the importation of the Chinese seeds and their grafting with the native plant was a "grave blunder." Crole, for instance, remarks that these foreign shrubs, with their small leaves and undersized stature, were entirely unsuitable for the ecological landscape of upper Assam. For him, the indigenous variety far surpassed its Chinese counterpart in output and quality, often fetching more than 10 to 20 percent per pound in the London market.[58] Crole suggests that the "successive delusions" of "impractical" scientific men – out of touch with Assam's *natural* capacities – were chiefly responsible for introducing this "curse" to the province. In an interesting semantic twist, Dr. J. Berry White of the Bengal Medical Service later called the *Thea bohea* a "miserable pest," a disastrous mistake that sired an unwanted hybrid and eclipsed the native Assam plant's true potential in the international market.[59] Indeed, Ukers goes a step further to argue that the China plant was inherently untransplantable and had produced similar results in Java and Ceylon (now Sri Lanka).[60] These insinuations notwithstanding, the early battles over tea and terroir highlight the growing ideological distance between "scientific" expertise and practical knowhow in these gardens that periodically emerged in the decades following. As the next chapter shows, a peculiar fiscal crisis in 1886 – induced by an overreaching government and its taxation policy – renewed this argument all over again. However, the eventual success of the British Assam tea plant, on its own, was an economic, nay *national*, godsend. In any case, by 1840, Dr. Lum Qua had died and the first contingent of Chinese workers and artisans deserted in droves, complaining of low wages, unhealthy climate, and a restrictive work environment. For our purposes, this period marks the beginning of the end of Chinese involvement in the Assam tea enterprise.

But the ship had already sailed for the Assam tea plant. Late in 1839, ninety-five tea chests arrived into London amidst growing imperial and

---

[58] Crole, *Tea: A Text Book of Tea Planting and Manufacture*, p. 25.
[59] Quoted in the *Journal of the Society of Arts* XXXV, November 19, 1886–November 11, 1887 (London: George Bell and Sons, 1887), p. 736.
[60] Ukers, *All About Tea*, p. 141.

nationalist fervor. As this consignment was put up for sale on March 17, 1840 by the EIC, numerous well-known tea merchants and brokers jumped in to claim a share. Among them were W. J. & H. Thompson, Joseph Travers & Sons, William & James Bland, and Messrs. Twining & Co. Though valued between 2s. 11d. and 3s. 3d. per pound, "patriotic buyers pushed prices up to between 8s. and 11d. per pound."[61] The general opinion, proffered by M/S Twining & Co., was decidedly upbeat:

> Upon the whole we think that the recent specimens are very favourable to the hope and expectation that Assam is capable of producing an article well suited to this market, and although at present the indications are chiefly in reference to teas adapted by their strong and useful flavour to general purposes, there seems no reason to doubt that increased experience in the culture and manufacture of tea in Assam may eventually approximate a portion of its produce to the finer descriptions which China has hitherto furnished.[62]

While these declarations were underway, more concrete commercial partnerships were being forged in London and Calcutta. With political appendages out of the way, Assam was now ready to welcome the rush of British speculators in service of this magic crop. Thus, by February 1838, a body of capitalists – both European and "native" – came together with a capital of 10 lakh rupees to form the Bengal Tea Association.[63] Mr. W. Prinsep, Secretary to the Association along with Mr. J. R. Colvin, Private Secretary to the Governor-General, entreated Lord Auckland to hand over the government experimental gardens to their guardianship. Meanwhile, on May 1, 1839, Mr. Prinsep formally requested the Revenue department, Government of India to take over the tea settlements in upper Assam and also secure the services of Mr. Bruce. A further moment of consolidation came when the Bengal Association, with help from Messrs. Cockrell & Co., and Boyd & Co., formed a union with another company that floated around the same time in London with a capital of £500,000.[64] Out of this emerged India's pioneer tea enterprise – the Assam Company – with a double board of directors, one each in London and Calcutta. It was virtually unchallenged till 1859 when a rival enterprise, the Jorehaut Tea Company was established. Nonetheless, the experimental phase lasted until about 1854, when the first respectable quantity – over a quarter of a million pounds – of Assam tea was successfully auctioned in London. These estates spread over a wide area of the province, stretching from the "lower Assam" districts

---

[61] Ukers, *All About Tea*, p. 147.   [62] Ibid.
[63] A lakh in the Indian numbering system equals 100,000.
[64] Barpujari, *Assam: In the Days of the Company*, p. 222.

of Goalpara and Kamrup in the west to the more abundant "upper Assam" tea-producing districts of Nowgong (now Nagoan), Darrang, Sibsagar, and Lakhimpur to the north and east. Collectively, these districts fell within what was termed the Brahmaputra valley, a rich alluvial floodplain drained (and periodically flooded) by the eponymous river. The crop was also produced in the flatter but also alluvial southern Surma valley, comprising the districts of Sylhet and Cachar, but bifurcated from its northern counterpart by the Khasi and Jaintiya hill range.[65] This study of tea refers to all these districts in some form or other. (See Map 0.2)

By 1858–59, close to 693,249 lbs. of Assam tea, produced at eighteen factories across these regions, were dispatched to the London market.[66] As the stage was set for the expansion of the tea trade and Assam's absorption into the networks of colonial capitalist economy and trade, the EIC surgeon John M'Cosh's words seemed to ring true: "Articles more precious than silver and gold grow wild upon [her] mountains."[67] In his introductory address to the Royal Medico-Botanical Society in London in 1839, quoted at the beginning of this chapter, Dr. G. G. Sigmond was emphatic. For him, these "discoveries" were not just "simple" cultivations; instead, they were to benefit the British Empire more than the "most brilliant discovery, or the most splendid achievement" till then.[68]

---

[65] For an extended description of Assam's geography, see John M'Cosh, *Topography of Assam* (Calcutta: Bengal Military Orphan Press, 1837); B. H. Baden-Powell, *The Land-Systems of British India: Being a Manual of the Land-Tenures and of the Systems of Land-Revenue Administration Prevalent in the Several Provinces*, Vol. III (Oxford: Clarendon Press, 1892); *Physical and Political Geography of the Province of Assam* (Calcutta: Assam Secretariat Printing Office, 1896); and Barpujari, *Assam: In the Days of the Company*, especially the Introduction.

[66] Quoted in Barpujari, *Assam: In the Days of the Company*, p. 227.

[67] M'Cosh, *Topography of Assam*, p. 31.

[68] Sigmond, *Tea: Its Effects, Medicinal and Moral*, p. 144.

2 Agriculture or Manufacture?

> Analysis of agricultural activities in narrow economic terms – as simply a matter of counting bushels and bales, of calculating profits and losses – obscures the more subtle relations between crop and cultivator.[1]

Ideological triumphalism and operational expediency were often difficult to reconcile. If Assam tea became the alibi for an elaborate British rhetoric of botanical mastery and agrarian "modernity" during the First Opium War, sustaining such positions in the face of tough market regulations was another matter. Nonetheless, for the planting community at least, ideological authority was not an abstraction – it emerged from their purported access to esoteric tea "science," "expert" horticultural knowhow, and direct agricultural involvement in the estates. These proclamations, and zealously guarded terrains of power, came under considerable scrutiny during a bitterly fought and long-drawn judicial battle between two contending parties – the Government of India (hereafter GOI) on the one hand, and planters and the Indian Tea Association (hereafter ITA) on the other. The war of words lasted from the mid-nineteenth till the early twentieth century, and came from an unlikely source: fiscal policy.

On March 19, 1918, it was officially decreed that income from tea plantations, hitherto exempt as "agricultural" produce, be taxed. While the storm clouds of taxation had been gathering for the enterprise since the 1860s, the Income Tax Act VII of 1918 was a definitive pronouncement. For the GOI, the argument was clear. This was an industry that used "modern" machinery, advanced techniques of manufacture, and mechanical processes to render the raw material "fit to be brought into the market."[2] No longer could plantations (in Assam in our case) claim immunity from the income tax by invoking specious claims to its agrarian

---

[1] See T. H. Breen, *Tobacco Culture: The Mentality of the Great Tidewater Planters on the Eve of Revolution* (Princeton, NJ: Princeton University Press, 1985), p. 21.
[2] This language was first used in the Income Tax Act of 1886; see W. H. Grimley, *An Income Tax Manual Being Act II of 1886, With Notes* (Calcutta: Thacker, Spink & Co., 1886), p. 7.

character. The ramifications, for administrators and planters, were enormous.

This chapter is not concerned with the economic dimension of the income tax debates per se, but the theoretical (indeed ideological) conundrum they precipitated: were plantations essentially manufacturing units, or were they agricultural in nature and operation? Could these aspects be separated and quantified? Conversely, did tea as *leaf* have no meaning – legally, materially, scientifically – distinct from its transformation as a finished product? Could the history of the plant and the plantation ever be divorced? Did ideology and practice ever come together in this commodity history, and under what constraints?

The tea income tax was, however, not about fiscal control. In fact, the economic import was only a very limited aspect of its overall history, and ambition. Instead, for the two parties involved, arguments for and against taxation were attempts to define the essential character and culture of the enterprise. It sought to identify and legitimize who had *natural* authority and expertise to do so, and why. The logics invoked for this purpose – discursive, political, and botanical – though connected, often led to contradictory conclusions. In the process, it also shows that the tea enterprise's self-sanctioned ideology of agrarian improvement was not just skin-deep and riven with inconsistencies; it could also be expeditiously jettisoned as and when required in the interest of profits and trade.

Before the Courtroom: "Natural" History and Plantation "Authority"

Good tea is made on the garden and not in the factory.[3]

What gave planters and the ITA discursive, material leverage and confidence to make their case? How could they argue so vehemently, and for so long, that the tea enterprise was an indivisible, organic operation? In this section, I show that though the income tax liability provided an expedient economic context for planters (and the ITA), their claims drew upon a long history of theory and praxis.[4]

---

[3] Harold H. Mann, *The Factors Which Determine the Quality of Tea*, Indian Tea Association Bulletin No. 4/1907, reprinted September 1913, p. 23, MSS EUR F/174/1515, Asian and African Studies, British Library, London.

[4] In addition to Lieutenant Colonel Money and David Crole's work invoked by Sir Mookerjee in his judgment as we shall see in the next section, planters would have been familiar with some of the following: G. D. Hope, *Memorandum on the Use of Artificial Manures on the Tea Estates of Assam and Bengal – Decade 1907–1917* (Calcutta: Star Printing Works, 1918); C. A. Bruce, *An Account of the Manufacture of the Black Tea, As Now Practiced at Suddeya in Upper Assam* (Calcutta: Bengal Military Orphan Press, 1838); M. Kelway Bamber, *A Text Book on the Chemistry and Agriculture of Tea Including the*

As discussed in the Introduction, an elaborate horticultural agenda and imperial geopolitics were intimately connected with the birth of the Indian tea enterprise. In its wake, a phalanx of botanists, scientist-administrators, itinerant travelers, institutions, and military men gave shape to this important article of trade in British India.[5] Though I do not intend to repeat that story here, it is imperative to revisit the agrarian context of tea's origins. A monoculture cash crop, it was allegedly found growing "wild" in India's northeastern frontier, and later took on an independent identity and character.[6] Admittedly, as we have seen earlier, there is a predicable telos to this textbook history – an unrefined shrub found in the "state of nature," and drunk by the "tribal" Singphos and

---

Growth and Manufacture (Calcutta: Law Publishing Press, 1893); Harold H. Mann, *The Tea Soils of Cachar and Sylhet* (Calcutta: The Indian Tea Association, 1903); A. F. Dowling, compiled, *Tea Notes* (Calcutta: D. M. Traill, 1885); C. H. Fielder, "On the Rise, Progress, and Future Prospects of Tea Cultivation in British India," *Journal of the Statistical Society of London*, 32, 1 (March 1869): 29–37; P. H. Carpenter and C. J. Harrison, *The Manufacture of Tea in North-East India* (Calcutta: The Indian Tea Association, 1927); W. T. Thiselton Dyer, *The Botanical Enterprise of the Empire* (London: Eyre and Spottiswoode, 1880); Samuel Baildon, *Tea in Assam: A Pamphlet on the Origin, Culture, and Manufacture of Tea in Assam* (Calcutta: W. Newman & Co., 1877); *Correspondence Regarding the Discovery of the Tea Plant of Assam*, Proceedings of the Agricultural Society of India (Calcutta: Star Press, 1841); H. A. Shipp, *Prize Essay on the Cultivation and Manufacture of Tea in Cachar* (Calcutta, 1865); J. W. Masters, "A Few Observations on Tea Culture," in *The Journal of the Agricultural and Horticultural Society of India*, Vol. III, Part I, January to December 1844 (Calcutta: Bishop's College Press, 1844); Spencer Bonsall, "Tea: Its Culture and Manufacture; With Directions for the Soil, Character of Climate, Etc., Etc., Adapted to the Culture of the Plant in the United States, From Practical Experience, Acquired by a Residence of Six Years in Assam," in *Report of the Commissioner of Patents for the Year 1860: Agriculture* (Washington, DC: Government Printing Office, 1861), 446–467, and *The Tea Cyclopaedia: Articles on Tea, Tea Science, Blights, Soils and Manures, Cultivation, Buildings, Manufacture Etc., With Tea Statistics* (London: W. B. Whittingham & Co., 1882); this is only a representative, and not an exhaustive, list. Despite referring mostly to Assam, these studies apply to the tea enterprise as a whole.

[5] Also see Jayeeta Sharma, *Empire's Garden: Assam and the Making of India* (Durham and London: Duke University Press, 2011), especially Part I; Jayeeta Sharma, "British Science, Chinese Skill and Assam Tea: Making Empire's Garden," *Indian Economic and Social History Review*, 43, 4 (2006): 429–455; Roy Moxham, *Tea: Addiction, Exploitation and Empire* (New York, NY: Carroll and Graf Publishers, 2004) and Bodhisatwa Kar, *Framing Assam: Plantation Capital, Metropolitan Knowledge and a Regime of Identities, 1790s–1930s*, Unpublished PhD Dissertation (New Delhi: Jawaharlal Nehru University, 2007).

[6] Laurelyn Whitt uses the term "biocolonialism" to refer to structures and methods of coercion – including biological transformation – used by the dominant culture to extract, appropriate, damage, and change "indigenous" ways of life, agrarian patterns, "knowledge and value systems," and environment. In this assessment, the introduction of "monoculture" farming and the attendant "undermining of plant genetic diversity" also refers to another form of biocolonialism; see Laurelyn Whitt, *Science, Colonialism, and Indigenous Peoples: The Cultural Politics of Law and Knowledge* (New York, NY: Cambridge University Press, 2009); this debate remains insufficiently theorized in terms of tea's advent in an otherwise wet-rice cultivating culture in Assam.

Khamtis of Assam waited for certification as marketable tea in the hands of metropolitan scientists, administrators and experts. Thus, despite early efforts to validate the "native" specimen – by the Bruce brothers (Robert and Charles) in 1824, and by Lieutenant Andrew Charlton in 1834 – we know that it was only later, "in the more propitious circumstances of the Tea Committee's workings" that the Assam plant received due certification.[7] Among others, the Tea Committee, established in February 1834 by Governor-General Bentinck, included the celebrated Danish-born botanist Nathaniel Wallich. In July 1835, Wallich, a geologist and surgeon-naturalist John M'Clelland, and another celebrated botanist William Griffith headed a "scientific delegation" to Assam to report on the earlier amateurish findings. Within this body of experts, the "young Turk" Griffith[8] famously pronounced that only by importing "Chinese seeds of unexceptionable quality" could the "savage" Assam plant be reclaimed as fine tea.[9] As this wisdom was unquestioningly accepted, Robert Fortune (with experience working in the Edinburgh Botanic Gardens) was dispatched to China in 1842 to send back tea seeds and live plants from the Celestial Kingdom. Alongside, G. J. Gordon was instructed by the Calcutta Botanic Gardens to "smuggle tea seeds out of China."[10] In this frenzy, Chinese tea-men, workers, and expert planters also made their way to upper Assam. The era of grafting the indigenous plant with its superior "oriental" cousin had begun in earnest. Being similar in plant bionomics, topographical origins and climatic zones, it was felt that this hybridization would yield handsome results. History, of course, would soon prove otherwise. As lands began to be cleared and colonized for these plantations, the hegemony of metropolitan science and technocratic expertise in tea cultivation – rather than the nuances of planting – continued to reign supreme. By 1865, however, the winds of despair began to set in, and the industry went into a crisis similar to the "South Sea Bubble."[11] Though numerous factors were to blame, the preponderance of esoteric knowledge over field experience had taken its toll.

---

[7] See Sharma, "British Science, Chinese Skill and Assam Tea: Making Empire's Garden," p. 439.
[8] On the scientific career of Wallich, and his intellectual run-ins with Griffith, see David Arnold, "Plant Capitalism and Company Science: The Indian Career of Nathaniel Wallich," *Modern Asian Studies*, 42, 5 (2008): 899–928.
[9] See Sharma, "British Science, Chinese Skill and Assam Tea: Making Empire's Garden," p. 440.
[10] Ibid., p. 442.
[11] On the tea boom, and its collapse, see Sir Percival Griffiths, *The History of the Indian Tea Industry*, especially chapters 6, 7, and 8; also see Rana Partap Behal. *One Hundred Years of Servitude: Political Economy of Tea Plantations in Colonial Assam*.

Agriculture or Manufacture?   53

An official report by commissioners in 1868 was scathing in its condemnation of the "rage for extending cultivation" since 1852 that led to the crash. Mincing no words, it blamed "extravagant" establishments, reckless expenditure, disregard for labor health, bad management, unsuitable locations, uncultivated "vacancies," and want of capital as responsible for the industry's dismal state of affairs.[12] More importantly, the report argues that in the wake of inexperience and ill advice, "great pressure had been put on Garden Managers ... to make up quantity without reference to quality."[13] For an enterprise that relied on taste as its ultimate hallmark, this was a serious charge. In a moment of disclosure, the unsuitable cultivation of the China plant in Assam is unambiguously marked as a primary cause of the fall in production: "the indigenous and hybrid varieties are superior, not only in strength, but also in productive power and facility in manufacture."[14] Reneging on an entire generation of expert opinion, this botanical oversight in the plantation's history was an embarrassment, at least for the government. Field personnel, especially planters, were far less generous with their criticisms of the Chinese experiment. In 1872, Lieutenant-Colonel Edward Money, with eleven years' experience in tea, published an award-winning essay on its cultivation and manufacture.[15] As this pamphlet ran into three editions, one message was clear – the indigenous species commanded higher prices, and produced a bolder flavor than its Chinese counterpart. Money was, however, despondent that a "pure specimen" of this original plant was now irrevocably lost, and all that remained was the hybrid *jat* or variety. In other words, no garden in the province was "wholly indigenous" or "wholly China" anymore. Looking back, Money rued that "great mischief" had been done by this "prejudicial" governmental policy of cross-pollination.[16] In his opinion, it was a missed opportunity not to let the Assam shrub stand on its own for: "the tea produced by it was as superior to China Tea as gold is to silver."[17] The moot point about Money's argument was that the reckless importation of this "foreign" seed – hardy and easy to grow – had swelled the size of plantations in

---

[12] See *Report of the Commissioners Appointed to Enquire into the State and Prospects of Tea Cultivation in Assam, Cachar and Sylhet* (Calcutta: Calcutta Central Press Company Ltd., 1868), especially pp. 12–18, ASA.
[13] Ibid., p. 14.   [14] Ibid., p. 15.
[15] Lieutenant-Colonel Edward Money, *The Cultivation and Manufacture of Tea*, third edition (London: W. B. Whittingham & Co., 1878).
[16] Ibid., pp. 48–51.
[17] Lieutenant-Colonel Edward Money, *The Tea Controversy (A Momentous Indian Question). Indian versus Chinese Teas. Which are Adulterated? Which are Better? With Many Facts About Both and the Secrets of the Trade* (London: W. B. Whittingham & Co., 1884), p. 7.

eastern India beyond manageable, and profitable, control. While acreage continued to increase, the finer and more essential points of the enterprise, namely the "class of plants" produced and quality, had been compromised.[18] In a later tract published in 1884 on this raging controversy, Lieutenant Money provided additional reasons for his earlier claim. Beyond the stated fine-points of the indigenous plant, other elements were introduced:

> Indian [tea] is superior to Chinese tea because ... [it] is grown and manufactured on ... estates under the superintendence of educated Englishmen, and skill and capital are combined to produce the best possible article. In China the Tea is produced, in many cases, round the cottages of the poorer classes, collected and manufactured in the rudest way ... Apart [from the difference mentioned above], and the far better climate for Tea India possesses, it is *natural* therefore that the Indian should be the better article.[19]

In this assessment, the natural brilliance of Assam tea resulted from a combination of factors – climatic, racial, and botanical. Channeled through the munificence of European agrarian superintendence, these disparate aspects of the plant's inherent goodness came together to produce the valuable commodity. Admittedly, this wisdom was an interesting and paradoxical reversal of the earlier "scientific orthodoxy."[20] Not only was the allegedly wild native plant superlative on its own, the beneficial hand of the English planter added to its worth and price. As Money wryly comments without naming anyone: "it is not strange that those who never saw Tea grown or manufactured, discussing what they are ignorant of, write nonsense."[21] David Crole, mentioned earlier, was even more blunt with his criticism of this "blunder." Writing from Chelsea, England in March 1897, Crole argued that the object lesson of this "folly" was to acknowledge that "a jury for such a matter composed entirely of men of science, without a leaven of practical men" should never have been convened.[22] Similarly, Dr. J. Berry White of the Bengal Medical Service called the *bohea* a "miserable pest," a disastrous mistake that sired an unwanted hybrid and eclipsed the native Assam plant's true potential in the international market.[23]

---

[18] Edward Money, *The Cultivation and Manufacture of Tea*, p. 48.
[19] Money, *The Tea Controversy*, p. 8; emphasis mine.
[20] I borrow the term "scientific orthodoxy" from Sharma, "British Science, Chinese Skill and Assam Tea: Making Empire's Garden," p. 442.
[21] Money, *The Tea Controversy*, p. 8.
[22] See David Crole, *Tea: A Text Book of Tea Planting and Manufacture* (London: Crosby Lockwood and Son, 1897), p. 25.
[23] Quoted in the *Journal of the Society of Arts*, Vol. XXXV, November 19, 1886–November 11, 1887 (London: George Bell and Sons, 1887), p. 736.

Agriculture or Manufacture? 55

It is historically inadequate to classify such opinion as "superficial [outbursts and] reading of theory versus practice by an interested party."[24] In the tea enterprise, policy, profits and performance were intimately – if expediently – connected. As it is, the botanical doxa of "practical men" such as Money, Crole and others were internalized by later generations of planters, tea-men, and estate agents in eastern India. For them, agrarian supervision was an indispensable, indeed indivisible *sine qua non* of a successful plantation. After all, without correct judgment about its organic ingredients – soil, climate, moisture, hydrology, leaf, and land – mechanical artifice and expert theorizing could do little to enhance natural yield and quality. In their view, techno-scientific "improvement" and mechanization neither sequestered nor supplanted these agricultural functions.[25]

Rhetorically speaking, it was not unusual to find planters refer to the garden as a "great machine" and themselves as its "prime mover."[26] A proprietorial arrangement between three players – the tea plant, the manager, and the estate – was at the center of this ostensible relationship.[27] The acronym P.G. or "practical gardener" was much in currency among tea-men during the early days. It was assumed that training in husbandry, English gardening, and nursery work naturally qualified one in this choice of career. After all, no less an authority than Sir William Turner Thiselton-Dyer had argued: "agriculture in the tropics is essentially extended gardening."[28] As the enterprise emerged from

---

[24] See Sharma, "British Science, Chinese Skill and Assam Tea: Making Empire's Garden," p. 441.

[25] See Chapter 3 for an analysis of the complicated relationship between scientific remedy and practical solutions in the Assam estates; also see Timothy Mitchell, *Rule of Experts: Egypt, Techno-Politics, Modernity* (Berkeley and London: University of California Press, 2002), especially Part I.

[26] See F. T. R. Deas, *The Young Tea Planter's Companion: A Practical Treatise on the Management of a Tea Garden in Assam* (London: S. Sonnenschein, Lowrey & Co., 1886), especially Part II; see also, Richard White, *The Organic Machine: The Remaking of the Columbia River* (New York, NY: Hill and Wang, 2005).

[27] Though skilled and unskilled "coolie" labor, "native" agents and intermediaries, accountants and European engineers, Marwari moneylenders, medical men and assistants formed integral, indeed indispensable actors of the estate, this chapter is not about them; see Sharma, *Empire's Garden*, Kar, *Framing Assam*; Amalendu Guha, *Planter Raj to Swaraj: Freedom Struggle and Electoral Politics in Assam, 1826–1947* (New Delhi: ICHR, 1977), and Rana Partap Behal, *One Hundred Years of Servitude* for an assessment of their histories in the Assam plantations.

[28] Quoted in *Bulletin of Miscellaneous Information*, Royal Botanic Gardens, Kew (London: His Majesty's Stationery Office, 1906), 97; Thiselton-Dyer was Kew's Director from 1885 to 1905; see also his, *The Botanical Enterprise of the Empire* (London: George E. Eyre and W. Spottiswoode, 1880); sometimes, the analogy was carried a little too far. In an elaborate Linnaean conceit drawn up in 1885, planters referred to themselves as the genus *Assamiensis* with five distinct sub-species – the young man with little capital; the Mincing Lane "expert"; the youth with connections; the practical engineer; and the P.G.

collapse, it dawned on field personnel that careful attention to all stages of crop cultivation and manufacture was necessary. Though labor remained a soft target in manipulating profits, these other processes assumed renewed significance. Writing in the journal of the Agricultural and Horticultural Society of India in 1844, J. W. Masters, Superintendent of the Assam Company, critiqued prevailing opinion that saw wanton tree felling, jungle clearing, and leaf plucking automatically leading to windfall gains. He wryly observed that "every bud that produces leaves, produces roots."[29] Put differently, Masters demonstrated that the enterprise's paying proposition – namely its crop – was intimately tied to methods of cultivation. He argued that an informed tea "culture" had to be instilled before estimates of profits and losses could be drawn up. Subsequent writers were to further expand on this poignant agrarian term. In an interesting report filed with the United States Patent office in 1861, Spencer Bonsall, a tea planter with six years' experience in Assam recommended that the article had a good chance of succeeding in the country's "Southern part."[30] After all it was an "object of agricultural produce, and [bore] a close resemblance to the vine. Skill and care, both in husbandry and preparation, are quite as necessary to the production of good tea as of good wine."[31] Bonsall also boasted about "making a machine" to roll leaves, though, by his admission, it could not give them the "proper spiral twist."[32] He was, however, emphatic that plucking needed human hands as no appliance could possibly be invented to distinguish old and young leaves.

What were these skills in husbandry that Bonsall referred to? Though details of tea cultivation and manufacture need not detain us here,[33] some assessment of the steps between the field and cup is necessary. In the dense forested regions of upper Assam, for instance, the process usually

---

or the "all round practical" man, quoted in *The Indian Planters' Gazette and Sporting News*, July 7, 1885, Microfilm Collection MFM.MC1159, Asian and African Studies, British Library, London.

[29] J. W. Masters, "A Few Observations on Tea Culture," in *Journal of the Agricultural and Horticultural Society of India*, Vol. III, Part I (January to December 1844), 4; the Assam Company, established in February, 1839 was the first joint-stock holding in the history of Indian tea but returned no profits till around 1859. For its "official" history, see H. A. Antrobus, *A History of the Assam Company 1839–1953* (Edinburgh: T. and A. Constable Ltd., 1957).

[30] Spencer Bonsall, "Tea: Its Culture and Manufacture: with Directions for the Soil, Character of Climate Etc., Etc., Adapted to the Culture of the Plant in the United States, from Practical Experience, Acquired by a Residence of Six Years in Assam," in *Report of the Commissioner of Patents for the Year 1860: Agriculture*, House of Representatives Papers, 36th Congress, 2nd Session, No. 48 (Washington, DC: Government Printing Office, 1861), p. 453.

[31] Ibid.   [32] Ibid., p. 464.

[33] See footnote 4 above for a list of works on this aspect of the enterprise.

began with jungle clearing, with care taken not to eliminate all the "natural" leguminous varieties such as the clover or gram. Being rich in nitrogen, this class of forests provided the necessary nutrient for tea plants almost by default. As firewood was cut and forests cleared, vigorous hoeing, forking, and hand weeding prepared the ground for the plantations. Meanwhile, tea seeds, either in nurseries or in clearly earmarked spaces within the estates, were sown for germinating. The tea flower usually appeared in the autumn, and another round of seeds was ready for use the following year. As seedlings appeared, transplanting them before the rains was a necessary step. Lieutenant Money observes that it was only by teaching, and then by practice, that either Europeans or "natives" could be made proficient in this delicate task.[34] He reminisced from experience that failure in transplanting was common and frustrating.

As the plant grew into maturity, pruning was the next important modus operandi in a garden. Pruning, or trimming was required both to induce flushes (or leaf) and for the overall health of the plant. As before, manuring and hoeing followed a fully pruned garden. The real test came the following "season." Succulent leaves (the proverbial two leaves and a bud) and hardy shoots, being the kernel of this enterprise, were the most prized possessions. Depending on locality, climate, soil, rainfall, and the absence of blights, the first flush usually appeared anytime between February and April, and continued till around the close of December. Tea men widely acknowledged that it took anywhere between two to three years for saplings to yield their first full flush. As leaves were plucked, the next stage in "rendering it fit for the market" began.[35]

The mechanical steps to transform the raw produce into the "dark-coloured, crisp tea of commerce" involved withering, rolling, fermentation, and firing (or drying).[36] Despite the inducement of various contrivances (automated and semi-automated) in tea manufacture, the vast "canon" on this subject consistently stress experience and judgment as critical ingredients. Lieutenant Money, for instance, argues that the best

---

[34] Edward Money, *The Cultivation and Manufacture of Tea*, p. 77.
[35] See the use of this phrase in the Indian Income Tax Acts, namely Act II of 1886, and Act VII of 1918 discussed in the previous section. To be sure, these demanding physical tasks were mostly carried out by estate labor in harsh and inflexible working conditions. Indeed, labor historians, especially of Assam, have argued that these elaborate processes of tea cultivation and manufacture necessitated a managerial and administrative "rationale for residential compulsion," in Behal, *One Hundred Years of Servitude*, especially chapter 2.
[36] Sifting, sorting, packing, and transporting the finished version come next, though I do not focus on these stages of tea manufacture in this book.

test for withering was the human ear: "fresh leaf squeezed in the hand, held near the ear, crackles, but no sound should be heard from withered leaf."[37] Crole contends that for "pucca" tea – his vernacular analogy for the perfect crop – over-withering and under-withering had to be prevented. He suggests that artificial means were mostly responsible for tea becoming over-shrunk and friable.[38] As M. Kelway Bamber, one-time chemist to the ITA, remarked: "there is no remedy when once the damage is done."[39] Rolling was the in-between step to prepare the withered leaf to exude its essential juices, and ready it for fermentation. Bamber describes fermentation, or oxidation as the "most important in the whole manufacturing process" as brew color, flavor, and quality depended on it.[40] The last step, firing, removed all moisture from the fermented leaf, ideally retained its essential oils and readied it for sifting, sorting, and eventual shipment to auction houses in Calcutta and Mincing Lane, London.

As mechanization increasingly took over the above manufacturing processes, especially after the 1880s, words of caution for managers and tea-men continued to be issued. In addition to weather conditions dictating the extent and nature of fermentation, care had to be taken to sanitize the fermenting room from "microbes."[41] Furthermore, galvanized iron plantings and zinc sheets used in this process, if not separated properly, left traces in the final tea. The mid-nineteenth century saw the advent of numerous tea-rolling machines (Kinmond's, Hayworth, Nelson, Jackson, and Cruickshank, among others). Despite reducing production time, these machines often broke freshly withered leaves due to the pressure applied at the start. Samuel Baildon notes: "what I have sometimes considered to be most important in machine-rolling, is the conduct of the engine-driver ... a man rolling leaf by hand goes very slowly to the work at first, but the nearer the leaf gets to the finish, the faster he rolls. Precisely the same rule should be followed by machinery."[42] Colonel Money argued that the tea-rolling machine merely *prepared* the leaf for the act, and did not supersede human intervention. As it were, for Money, these "machines do not give the nice final twist which is obtained by the hand."[43] For Bamber, these rolling machines reminded him "of action

---

[37] Money, *The Cultivation and Manufacture of Tea*, pp. 120–121.
[38] Crole, *Tea: A Text Book of Tea Planting and Manufacture*, p. 139.
[39] See Bamber, *A Text Book on the Chemistry and Agriculture of Tea including the Growth and Manufacture*, pp. 221–222.
[40] Ibid., p. 225.
[41] Harold H. Mann, *The Factors Which Determine the Quality of Tea*, p. 25, MSS EUR F/174/1515, Asian and African Studies, British Library, London.
[42] Samuel Baildon, *Tea in Assam: A Pamphlet on the Origins, Culture, and Manufacture of Tea in Assam* (Calcutta: W. Newman & Co., 1877), p. 29.
[43] Money, *The Cultivation and Manufacture of Tea*, p. 116.

Agriculture or Manufacture? 59

similar to the movement of the hands of the coolies."[44] Gibbs, Victoria, Duplex, Davidson, and Sirocco machines were the acknowledged workhorses of tea drying. Crole was categorical that though automated machines satisfactorily achieved the first two stages of the manufacturing process with adequate supervision, firing required a "great deal of nicety" for human control to be relinquished.[45] Excessive heat irrevocably burnt rolled leaves, and depleted caffeine. P. H. Carpenter and C. J. Harrison of the ITA opined in 1927 that caffeine was one of tea's chief characteristics, and its loss during the firing stage, though common, could be prevented by temperature control.[46] Harold Mann rued: "in all of the automatic machines at present on the market, the leaf is more or less stewed in the early part of the operation, the ideal conditions of firing do not at present seem altogether attainable."[47] Baildon was more direct, though dramatic, when he suggested that the sun induced better tea drying that any appliance, automated or otherwise.[48] Bamber reminded managers that it was at this stage that a great deal of tea "is most likely to be spoiled, either by over or under firing, unless continued supervision is given."[49] David Foulis, one-time planter in the Cachar district of Assam, noted in his unpublished diary: "it is not by windy letter writing and elaborate accounts that tea will be made to pay, not by office work but by the personal interest, constant attention ... to the most trifling details of the work."[50]

Armed with these "experts" and field-sanctioned knowledge, planters (and the ITA) set out to claim *natural* and indivisible authority over the plant and its production space. The income-tax debates that ensued between the colonial government and the planting community revealed much more than a fiscal impasse. To these we now turn.

---

[44] Bamber, *A Text Book on the Chemistry and Agriculture of Tea including the Growth and Manufacture*, p. 222.
[45] Crole, *Tea: A Text Book of Tea Planting and Manufacture*, p. 150.
[46] P. H. Carpenter and C. J. Harrison, *The Manufacture of Tea in North-East India*, the Indian Tea Association (Calcutta: Catholic Orphan Press, 1927), pp. 3–4.
[47] Mann, *The Factors Which Determine the Quality of Tea*, p. 28.
[48] Baildon, *Tea in Assam*, p. 36.
[49] Bamber, *A Text Book on the Chemistry and Agriculture of Tea*, p. 235; to be sure, machinery was also widely used in the next stages, namely sifting, sorting, mixing, and packing. As with the previous functions, opinion differed regarding the *extent* of its use and efficacy. Colonel Money, for instance, did not "believe in *any* machine for Tea sifting, simply because it is not a regular process"; see Money, *The Cultivation and Manufacture of Tea*, p. 132, emphasis in the original; see also, Claud Bald, *Indian Tea: Its Culture and Manufacture, Being a Textbook on the Culture and Manufacture of Tea*, second edition (Calcutta: Thacker, Spink and Co., 1908).
[50] Journal of D. Foulis, "The Tea Assistant in Cachar," MS 9659, National Library of Scotland Manuscript Collection.

To Tax or Not to Tax

> To the tea planter, who works for so many hours at the cultivation of his estate, who watches the daily growth of his crop with hope and anxiety, and who spends many hours of his leisure in the study of soils, of problems of manuring or draining, and of the progress of agricultural science, the idea that the tea industry is not essentially agriculture seems quite absurd.
>
> (Proceedings of the ITA, January 8, 1919)[51]

The plantation sector in British India had been traditionally kept outside the purview of direct taxation. As an agricultural concern, it paid land revenue to the government and a double impost, it was argued, would be both uncalled for and unjust.[52] Also, an additional income tax contravened Cornwallis' 1793 injunction, especially in permanently settled districts.[53] Thus, Act II of 1886 ruled in section 5 that:

Nothing [...] shall render liable to the tax –

(a) any rent or revenue derived from land which is used for agricultural purposes and is either assessed to land-revenue or subject to a local rate assessed and collected by officials of the Government, as such; or
(b) any income derived from –
　(i) agriculture, or
　(ii) the performance by a cultivator or receiver of rent-in-kind of any process ordinarily employed by a cultivator or receiver of

---

[51] Quoted in *The Indian Planters' Gazette and Sporting News*, January 25, 1919, Microfilm Collection MFM.MC1159, Asian and African Studies, British Library, London.

[52] See Sir Percival Griffiths, *The History of the Indian Tea Industry* (London: Weidenfeld and Nicolson, 1967), p. 557. In Assam, tea lands paid revenue under rules specifically enacted to stimulate European settlement and colonization in the province. The Wasteland Rules of 1838, the Old Assam Rules of 1854, the Fee-Simple Rules of 1862, the Leasehold Rules of 1876, and the Assam Settlement Rules 1886, which lasted well into the second decades of the twentieth century designated the purchase price, acreage grants, and revenue assessment policy for areas brought under tea cultivation. Ostensibly open to all, the revenue requirements under these rules virtually excluded local participation and capital investment in large-scale plantation enterprises till around the 1920; I elaborate on the agro-economic ramifications of these land tenure policies in Chapter 5.

[53] Cited in Griffiths, *The History of the Indian Tea Industry*, p. 557. Lord Cornwallis' proclamation of March 22, 1793, otherwise known as the Permanent Settlement, envisioned fixing the revenue assessment from landlords to the state in "perpetuity," thereby shielding them from price fluctuations while providing security of ownership. In real terms, however, this physiocratic doctrine of agrarian "improvement" had disastrous consequences; Ranajit Guha's *A Rule of Property for Bengal: An Essay on the Idea of Permanent Settlement* (Durham and London: Duke University Press, rpt. 1996) remains the best exposition of its history, intellectual antecedents, and fallout. See also S. M. Pagar, *The Indian Income Tax: Its History, Theory, and Practice* (Baroda, 1920).

rent-in-kind to render the produce raised or received by him fit to be taken to market, or

(iii) the sale by a cultivator or receiver of rent-in-kind of the produce raised or received by him, when he does not keep a shop or stall for the sale of such produce.[54]

While reiterating the earlier logic of exempting agriculture from taxation, this Act unwittingly opened up a Pandora's box of questions and paradoxes for the tea (and indigo) industry among other similar ventures. Were tea planters liable to the tax? Could the enterprise be apportioned between its agrarian and non-agrarian units? For now, law seemed to have the answers.

Amidst a hail of protests from planters and the tea association against the tax, the Advocate-General of Bengal, Sir G. C. Paul, ruled that tea (and indigo) planters were "producers" of an agricultural item and were therefore exempt.[55] Employees of the enterprise, drawing regular salaries and benefits, were not. Similarly, any "building owned and occupied in the immediate vicinity of the land" or "homestead appurtenant" to the land required for the operation and "necessary for the vocation" mentioned under clause (b) above were also exempt. For any other houses, the enterprise was liable to pay an impost.[56] The percentage of income from these portions was, however, not fixed. In trying to separate the seemingly indivisible, Sir Paul's interpretation only created further incredulity. As a newspaper report of May 11, 1886 emphatically stated:

The idea of exempting tea gardens and indigo factory managers, and taxing assistants is neither logical nor does it bear with it even the semblance of sanity.[57]

But under what pretexts were planters *producers* of agricultural goods in its essential, natural form? Under what logics were they cultivators untainted by the mechanized aspects of the plantation process? For the moment, further legal quibbling seemed to provide some analytic answers. Thus, replying to a charge that indigo planters in Bihar should be assessed on the income tax, Messrs. Woodroffe and Evans (hereafter W & E) argued otherwise. Though specific to indigo, their opinion of June 30, 1886 had direct ramifications for the tea enterprise.[58] For one, W & E sought

---

[54] See W. H. Grimley, *An Income Tax Manual Being Act II of 1886, With Notes*, pp. 6–7.
[55] Ibid., p. 7. [56] Ibid., pp. 7–8.
[57] *The Indian Planters' Gazette and Sporting News*, May 11, 1886, Microfilm Collection MFM.MC1159, Asian and African Studies, British Library, London.
[58] See judgment of Messrs. Woodroffe and Evans, "The Liability of Indigo Planters to Assessment of Income Tax," reprinted in *The Indian Planters' Gazette and Sporting News*, July 27, 1886, pp. 82–83.

to define "agriculture" and "cultivator" that had been left ambiguous in the 1886 Act. It did so by highlighting the *relationship* between the planter and the crop, the cultivator and the product that had only been imputed in the aforementioned legislation. For W & E, what distinguished the planter from the end-user was the investment of "risk" in a highly "precarious" and "peculiar" crop whose success depended on variables such as land quality, weather, climate, labor, skill, and supervision. In other words, the relationship of husbandry between a planter and the planted was purportedly an *organic* one, a necessary and indivisible step in exacting the value inherent in tea or indigo. If mechanized processes were used in transforming this to a "marketable" form, it is because the crop in its raw, biological state had no worth except to the planter. W & E thus forestalled attempts to deconstruct these two processes as separate – one agricultural and the other industrial. It argued against the logic of the 1886 Act by suggesting that:

If it were possible (which it is not) to distinguish between the income derivable by [the indigo planter] from the plant and the income derivable from the colouring matter, when extracted therefrom by the ordinary and necessary process, the latter income would also be exempt under section 5 (b) (ii) of the Act ... But we are of the opinion that [this section] may be left out of consideration, because there is no income separately attributable to the performance of any process in this case.[59]

W & E could easily have been referring to tea in this case. Predictably, planters in Assam quoted their judgment in arguing for tax exemptions. Accusing the GOI of hastily putting together provisions of this act, planters rued the "ill-considered" and "unbusiness-like blunders" of separating these domains of tea production for purposes of the impost.[60] As one report exasperatedly announced: "either we are agriculturalists, or we are not!"[61] Even murmurs of bringing a test case in the law courts to summarily settle the matter were heard. The GOI was, however, reluctant to proceed along these lines,[62] and the enterprise was largely kept outside the income tax till the second decades of the twentieth century. For now, the planters' point of view had been vindicated: the enterprise was essentially an agrarian venture, superintended under his close, involved watch. Machinery employed in the manufacturing process did not alter this organic bond between cultivator and crop.

---

[59] Ibid., p. 83; for a recent study of indigo in British India, see Prakash Kumar, *Indigo Plantations and Science in Colonial India* (Cambridge and New York, NY: Cambridge University Press, 2012).
[60] *The Indian Planters' Gazette and Sporting News*, September 14, 1886, p. 277.
[61] Ibid., September 7, 1886, p. 248.
[62] Sir Percival Griffiths, *The History of the Indian Tea Industry*, p. 558.

Agriculture or Manufacture? 63

This was a strange, almost paradoxical legal vision. Thus, while other employees of the tea industry were liable to the tax (thus pairing them as non-agriculturalists), planters were not. For law, it did not matter if assistants, engineers, and accountants simultaneously aided in transforming tea to its marketable state. Their services were avowedly mechanical in character and lacked horticultural immediacy needed for exemption. Theoretically, the 1886 act left the enterprise a divided and divisible house.

All was quiet for the next three decades or so. But as war clouds gathered on the horizon, the local governments of Bihar and Orissa declared in 1914 that sugar industries in these two regions were liable to the tax and were not exempt. The GOI agreed, and contended that the 1886 act had been erroneously applied in this case. In making its case, the government (and its legal counsel) introduced an ontological premise – that of "crude" agriculture vis-à-vis "modern manufacturing processes" – in this debate. In other words, the GOI argued that section 5 of the 1886 legislation had only been intended for basic "domestic processes" and not for sugar factories employing "up-to-date lines for the express purpose of making a business."[63] Alongside, the Government of Bengal (hereafter GOB) solicited the opinion of the then Advocate-General G. H. B. Kenrick on similar provisions for the tea industry. As the old feud rekindled, the powerful ITA and planters were once again at the center of a power struggle. But Kenrick was categorical in his verdict of December 1915:

> In my opinion ... [Section 5 of the 1886 Act] contemplates any simple process which is ordinarily undertaken by a cultivator to prepare his agricultural produce for market, but does not contemplate the complex and developed process of manufacture or preparation in modern factories equipped with machinery ... a tea factory containing modern scientific machinery or appliances for the preparation of tea for the market [is] liable to be taxed as income accruing and arising in British India.[64]

As anticipated, this did not go down well with either the ITA or the planting community. A war of words commenced between them and the government that lasted for the next six years. Though it was not until March 19, 1918 that a new legislation, the Income Tax Act

---

[63] Quoted in letter from Secretary to the GOI (Finance Department) to Secretary, ITA, in *The Indian Planters' Gazette and Sporting News*, August 3, 1918, pp. 115–116, National Agricultural Library (hereafter NAL), United States Department of Agriculture (USDA), Beltsville, MD.
[64] Ibid., p. 116.

64   Tea Environments and Plantation Culture

Figure 2.1 Cachar and Sylhet Tea Factory, Barak Valley, Assam, India, 1910, unattributed postcard © Mary Evans Picture Library

VII, was passed into law, the intervening debate is historically interesting.

For one, it introduced the language and parameter of "modernity" in making the manufactured portion of the tea enterprise liable for taxation. In real terms, as we shall see, this was a much more slippery terrain to demarcate. And it was not simply an economic matter. For planters and the ITA, agreeing to Kenrick's position implied jettisoning an elaborate operational arrangement that centered on tea as an organic substance. As mentioned above, this was not an agrarian idea alone: the counter-charge to GOI's position drew upon a long discursive, "scientific" and botanical wisdom on the subject. These protestations were also intimately tied to planter notions of legitimate authority and power – primarily over the plant, but also over its production space. But the paradox was hard to miss. The tea enterprise had long championed itself as the harbinger of social, material progress in an otherwise "backward" region of empire; contravening this ideological position by foregrounding its essential agricultural character was a practical, if not theoretical double-bind. The planters attempted to circumvent this irony by highlighting the organic indivisibility of the *Camellia sinensis*, or the tea plant, between its so-called "natural" (agrarian) and "manufactured" (business) parts. Ostensibly a financial issue, the income tax debates between the GOI

Agriculture or Manufacture? 65

and the tea enterprise in the early twentieth century was simultaneously an ideological, botanical, tug-of-war.[65]

Meanwhile, on March 27, 1918, the Assam branch of the ITA received a letter from the administration asking for help in assessing the "industrial" portion of the industry's income.[66] As the war waged on outside, a bitter, and oftentimes virulent, battle ensued between the plantation sector and the government for the next year-and-a-half. This letter to the Assam branch of the tea association awakened angry members that the GOI had already passed the new, revised income tax provisions into law. Accusing the latter of "under-hand" and "back-door" policies in implementing this new impost, the ITA and planters prepared to contest the 1918 Act tooth and nail.[67] For one, they argued that this new "inequitable" and "unjust" tax had been thrust on the industry without warning or intimation. Secondly, it was contended that the word "factory," seen as a necessary extension of the cultivator's agrarian functions in the 1886 Act, was expunged from Section 5 (1) (c) of the new legislation with the sole purpose of targeting the tea enterprise. Lastly, planters and the ITA ridiculed Section 43 (2) of the 1918 Act that expressly gave the Governor General in Council powers to instruct local governments to apportion between "agricultural income" and "business income" with regards to cases when these were in doubt.[68] In this, the ITA was not totally without local support. The Assamese planter and Congress councilor Ghanashyam Barua lobbied in the Imperial Legislative Council in support of his British counterpart's argument in this regard.[69]

On their end, the GOI was categorical that the new impost was not a "fresh" imposition on the enterprise, but "exaction" of an old tax erroneously un-levied. It further clarified that the word "factory" was excised due to its incompatibility in a section dealing with agriculture and with a view to removing "misapprehensions on the part of the assesses

---

[65] On this topic, also see Prakash Kumar, "Plantation Indigo and Synthetic Indigo: European Planters and the Redefinition of a Colonial Commodity," *Comparative Studies in Society and History*, 58.2 (April 2016): 407–431; Kristin Adal, "The Problematic Nature of Nature: The Post-Constructivist Challenge to Environmental History," *History and Theory*, Theme Issue 42 (December 2003): 60–74; William Cronon, *Nature's Metropolis: Chicago and the Great West* (Chicago, IL: W. W. Norton, 1991); and Leo Marx, *Machine in the Garden: Technology and the Pastoral Ideal in America* (New York, NY: Oxford University Press, 1964).

[66] See *The Indian Planters' Gazette and Sporting News*, April 27, 1918, pp. 477–478, NAL, Beltsville, MD.

[67] Ibid., May 18, 1918, p. 557, NAL.

[68] For the full Act VII of 1918, see *The Unrepealed General Acts of the Governor General in Council: From Act I of 1914 to Act XI of 1919*, Vol. VIII (Calcutta: Superintendent of Government Printing, 1919), pp. 250–274.

[69] Quoted in Amalendu Guha, *Planter Raj to Swaraj: Freedom Struggle and Electoral Politics in Assam, 1826–1947* (New Delhi: ICHR, 1977), p. 70.

in such *industries* as the tea *industry* as to their exact position."[70] Mincing no words, the GOI claimed that the jute industry, similar in character to tea, had lost its tax privileges bestowed prior to the war and that the latter should follow suit. It clarified that assessments under agricultural income would be made *only if* two conditions were met – "first, the whole of the raw material of his business must be raised from land of which he is the cultivator or received by him as rent in kind, and, secondly, he must employ no process to make the agricultural produce merchantable other than that ordinarily used by a cultivator."[71] In fact, Sir William Meyer (1860–1922), Finance Member of the GOI had already cautioned in 1916–17 that in addition to the export duty on tea (and jute), additional taxation could be levied if needed:

> Both these industries have been specially prosperous during the war and in the case of jute I think there is already a considerable consensus of opinion that if the financial situation created by the war should necessitate heavy additional taxation this is one of the first articles which might legitimately be taxed. The case of tea is hardly less strong, especially in view of the fact that in spite of its largely industrial character the tea business has for thirty years been exempted from income tax.[72]

But the last had definitely not been heard of the matter. This new characterization of the enterprise as industrial struck at the heart of how planters (and their overlords, the ITA) saw themselves. After all, as we have seen, for more than seven decades after tea experiments began in earnest in Assam, planters, field personnel, tea experts, and scientists (that included entomologists, botanists, and chemists) had harped on expert cultivation as key to the plant's (and by extension the plantation's) success in the international market. Jettisoning this essential trait was akin to an ideological and economic defeat.

The renewed protest against the government's policy after May 1918 thus critiqued the "simplistic" and "lame" logic of the new act. In a rhetorical move, the ITA questioned the GOI's definition of "modernity," and wondered if machinery could indeed be separated between its ancient and modern parts. It lampooned the government's reasoning by asking if a tea-estate's tax liability ended by employing equipment of an "antiquated" type. More to the point, planters and the ITA questioned: "does the Government require us to adopt [...] the method of ancient

---

[70] See letter from the Secretary to the GOI (Finance Department) to the Secretary, ITA, reprinted in *The Indian Planters' Gazette and Sporting News*, August 3, 1918, pp. 115–118, NAL; emphasis mine.
[71] Quoted in *The Indian Planters' Gazette and Sporting News*, May 18, 1918, p. 555.
[72] Quoted by the Secretary to the GOI (Finance Department) to the Secretary, ITA, reprinted in *The Indian Planters' Gazette and Sporting News*, August 3, 1918, p. 117, NAL, Beltsville, MD.

China?"[73] As this last point rekindled old debates on tea cultivation practices in eastern India, and the contentious use of Chinese hybrids,[74] the moot point was to now argue that the enterprise could not be distinguished in terms of its agrarian and industrial character. Interestingly, though predictably, the Bengal Mahajan Sabha also joined in the debate and sided with the anti-government position around June 1918. In a letter to the Secretary to the Finance department of the GOB, the association contended that despite the use of complex machines in a tea estate, "it cannot be argued by any stretch of reasoning that tea leaves are converted into a different material after undergoing the process of preservation."[75] To further buttress its case, the Sabha illustrated the case of jute and rice undergoing similar mechanized processes while retaining its essential agrarian trait. It cautioned the government that taxing the tea enterprise would set a worrying precedent for other agriculturists. An extended report by a special tea commissioner of June 7, 1918 was even more blunt. It reminded the government that factors regulating tea production varied from one district to the next, and a uniform policy of apportioning taxable income would be futile, if not foolhardy. Moreover, it argued that the major portion of the industry's profits was earned on the cultivation – the main elements existing in the leaf before it is brought to the factory.[76] In other words, the commissioner suggested that even if the 1918 Act was to be implemented in letter and spirit, and these two parts divorced, the tea enterprise's earnings from its industrial segment was "so infinitesimal a value that it would not be worthwhile taxing it" at all.[77]

To be sure, the anti-government stance was not about legal logics alone. In the face of stiff international competition from Java, Ceylon (now Sri Lanka) and China, planters and the ITA protested the ruinous economic doctrine of taxing income along with the current export and import duty. More importantly, it pointed out that the government's line of reasoning contravened, and contradicted, those of the Report of the Indian Industrial Commission (hereafter RIIC), also tabled in 1918:

> Improvement [of] agriculture was necessary, not only because it forms the basis on which almost all Indian industries must depend, but also for the further reason that the extension among the people of a knowledge of improved agricultural methods, and, in particular, of the use of power or hand-driven machinery, will benefit agriculturalists both by adding to their income and by its educative

---

[73] *The Indian Planters' Gazette and Sporting News*, May 18, 1918, p. 555, NAL.
[74] See the previous section for an elaboration of this point.
[75] See *The Indian Planters' Gazette and Sporting News*, June 8, 1918, p. 635, NAL.
[76] See "A Surprise Attack. And a Vigorous Counter Offensive," in ibid., p. 635, NAL.
[77] Ibid., p. 635, NAL.

effect... from the point of view of Indian Industrial development, the necessity for increased efforts for the improvement of agriculture is clear.[78]

While the timing of the 1918 Income Tax Act and the RIIC was serendipitous, the Industrial Commission Report strengthened the opposition's claim against the government's move. In a strongly worded section titled "Reform Penalized," planters and the ITA argued that the new tax regime was "retrogressive" and inconsistent with its avowed proclamations elsewhere.[79] By taxing business income, and exempting agriculture, the 1918 tax act created an iniquitous and artificial breach between these two processes of national growth. As one report wryly asked later: "Is agriculture not a business?"[80] Using the alibi of the RIIC, planters and the ITA targeted the new tax policy for "penalizing improved agricultural methods... and setting back the clock of agricultural progress in the country."[81] For them, the argument had come full circle – the industrial and agrarian portions of the tea enterprise were not only indivisible; they were also very much in line with the enlightened wisdom of the government's own body of experts. The paradox was obvious.

Judgment Day

Faced with an unprecedented barrage of protests and litigations, the GOI finally agreed to "an act of grace"[82] – it allowed the tea industry an extended exemption till March 31, 1919. As the deadline passed, the controversy failed to die down. Flustered and out of options, the government mulled a test case in the Calcutta High Court under Section 51 of the 1918 Income Tax Act to definitively settle the matter. Under this section, the "Chief Revenue authority" was empowered to bring to court "any questions [of assessment] arising from an interpretation of [any] of its provisions."[83] If executive deliberations had failed to come to a conclusion, the GOI felt that law, perhaps, had the final solution. Of course, a judicial review and ruling would hold true for the entire tea community. Accordingly, the Killing Valley Tea Company (hereafter KVTC) in Nowgong, Assam under

[78] See *Report of the Indian Industrial Commission, 1916–1918* (London: His Majesty's Stationery Office, 1919), especially pp. 52–57, Command Papers 51, House of Commons.
[79] See *The Indian Planters' Gazette and Sporting News*, August 3, 1918, p. 109, NAL.
[80] *The Indian Planters' Gazette and Sporting News*, January 25, 1919, p. 104, NAL.
[81] Ibid., August 3, 1918, p. 109, NAL.
[82] Vide, letter from the Secretary to the GOI (Finance Department) to the Secretary, ITA, reprinted in *The Indian Planters' Gazette and Sporting News*, August 3, 1918, p. 118, NAL.
[83] See *The Unrepealed General Acts of the Governor General in Council: From Act I of 1914 to Act XI of 1919*, Vol. VIII, p. 272.

the Calcutta managing house, James Finlay and Company, was chosen. A bench headed by acting Chief Justice Sir Asutosh Mookerjee, J Fletcher and J Chaudhuri heard *Killing Valley Tea Company Ltd. Versus The Secretary of State for India* and pronounced their judgment on May 31, 1920.[84]

KVTC was a sprawling estate of around 553 acres with a working capital of around 1 lakh rupees. Additionally, it was valued at around 77,000 rupees and employed "a European manager, an assistant manager, and around 800 'coolies'."[85] It was also reported that machinery was widely used by the Company to manufacture tea for export. Under these circumstances, the GOI solicited the Calcutta High Court's verdict on one central aspect of the decades-old debate: did the KVTC's operations and profits accrue from "agricultural" pursuits and therefore remain exempt from the income tax? In other words, the bulk of the legal burden was to identify whether "agricultural income" – as category and practice – existed at all in the tea industry that regularly used advanced mechanical processes. The statements, for and against, in this case are worth noting. On behalf of KVTC, their counsel argued:

After the tea bush has been planted and has arrived at a proper stage of maturity, the young green leaf is selected and plucked *by hand* from the bush – It is then dried or withered and rolled, dried and stored. The actual dried and rolled leaf, the produce of the tea bush, is then sent to the market. In the very early days of tea cultivation, the green leaf was dried or withered in the sun and was then rolled by hand. This *primitive* method was replaced by machinery. The effect of these processes being carried out by machinery in no way alters the processes or affects the results. It only leads to a quicker manipulation of the leaf ... the actual leaf of the tea plant, without the addition thereto of the processes above described, is of no value as a market commodity.[86]

The Crown's counsel disagreed:

It is contended on behalf of the Government that the manufacturing processes carried out in a *modern* tea factory, with scientific appliances and up-to-date machinery, are different from those ordinarily employed by a cultivator to render the produce raised by him fit to be taken to market. In former times, the process of manufacturing tea was *simple* and primitive ... the process employed in a *modern* tea factory goes far beyond this.[87]

---

[84] No. L.S. 83/8/6/20 in the High Court of Calcutta, see *Indian Law Reports* (hereafter ILR), Calcutta Series, Vol. XLVIII, January to December 1921 (Calcutta: Bengal Secretariat Legislative Department), pp. 161–176.
[85] Ibid., p. 161.
[86] Vide, *ILR*, Calcutta Series, Vol. XLVIII, January to December 1921, p. 169; emphasis mine.
[87] Ibid., pp. 169–170; emphasis mine.

While the GOI re-invoked the tired logic of "modernity" as its clincher, KVTC's deposition rehashed Woodroffe and Evans' perspective discussed earlier. In the face of old arguments, the Court sided with neither. Seeking closure to legal, if not botanical circumlocution, Sir Mookerjee was emphatic: "we are of the opinion that both the contentions are erroneous."[88] The Calcutta High Court decided that the moot point in the entire debate was one of apportionment. In other words, the tea enterprise was neither *wholly* agricultural, nor *wholly* industrial. If this judgment smacked of earlier wisdom, especially Sir G C Paul's after the 1886 Act, it was because this issue of divisible income had never been effectively settled. In this case, more than three decades after that original impost, the question of separating profits was once again made primary. Sir Mookerjee relied on, and referred to the textual authority of celebrated tea men – namely Lieutenant-Colonel Edward Money and David Crole – to argue that the "earlier part of the operation" of tea plantations, when tea bush is planted and leaf plucked, could be deemed agricultural.[89] However, the Court maintained that the "latter part of the process" where machinery was used to render the tea leaf into "marketable commodity" could not be termed thus "without violence to language."[90] In other words, law refused to agree to the Company's contention that these two processes are indivisible; for them, they were already always distinct. As the Court ruled:

There can be no doubt, in our opinion, that the entire process is a combination of agriculture and manufacture.[91]

In making its case for divisible income, the High Court referred to numerous precedents. The judgment in *Inland Revenue Commissioners v. Ransom* (1918 2 K.B. 709) was cited, where the company – chemists and manufacturers of medicinal herbs – was assessed excess profits duty on its first element. As far as law was concerned, it saw no reason "why a corporation, any less than an individual, should not be engaged in more than one trade or business at the same time."[92] It quashed the Company's

---

[88] Ibid., p. 169.
[89] Among Lieutenant Colonel's Money's award-winning treatises were, *The Cultivation and Manufacture of Tea* (London: W. B. Whittingham & Co., 1878). This "essay" was awarded the gold medal by the Agricultural and Horticultural Society of India in 1872. See also his, *The Tea Controversy (A Momentous Indian Question). Indian versus Chinese Teas. Which are Adulterated? Which are Better? With Many Facts About Both and the Secrets of the Trade* (London: W. B. Whittingham & Co., 1884). David Crole, an old tea hand, was once the manager of the Jokai tea estate in Assam; he is known for his *Tea: A Text Book of Tea Planting and Manufacture* (London: Crosby Lockwood and Son, 1897).
[90] Vide, *ILR*, Calcutta Series, Vol. XLVIII, January to December 1921, p. 170.
[91] Ibid., p. 171.   [92] Ibid., p. 172.

Agriculture or Manufacture? 71

claim that the raw tea plant was of no value commercially and needed to pass through the industrial stages to render it saleable:

The green leaf is not a marketable commodity for immediate use as an article of food, but it is a marketable commodity to be manufactured by people who possess the requisite machinery into tea fit for human consumption.[93]

Readers will note that this was a decidedly Marxian argument – while KVTC stressed that tea (as leaf) had no "use-value" to them, the Court opined that its "use-value" was an inherent property of the raw material that transformed it into an article of exchange for KVTC through mechanization. As Marx argued: "for the owner, his commodity possesses no direct use-value. Otherwise, he would not bring it to the market. It has use-value for others; but for himself its only direct use-value is as a bearer of exchange-value, and consequently, a means of exchange."[94] In essence, the Court's verdict was to tax the "exchange-value" component of the commodity (tea) while KVTC and planters claimed all along that its "use-value" and "exchange-value" were both the *result* of agricultural expertise. Nonetheless, despite the above verdict, the Court did not specify *where* the agrarian operations ended and manufacturing began in the tea-making process. In other words, as of May 31, 1920, the quantum of income from these two purportedly separable stages remained ambiguous and unresolved. In fact, Sir Mookerjee stated that no rules were then available to apportion these two incomes:

When they are framed and operate as statutory rules, an assessment may be made on such portion of the profits of the Company as do not fall within the description of "agricultural income."[95]

As is evident, *Killing Valley Tea Company Ltd. Versus The Secretary of State for India*, though mandatory for all tea establishments to follow, did not really settle the matter. The ITA was quick to suggest that such a "half-hearted judgment" could not stand for very long.[96] Accusing the legal mind of being incapable of understanding the "nice points in tea production," it presciently remarked that the ruling was an invitation for endless manipulation and disputes.[97] On its part, the Company's counsel, Sir B. C. Mitter argued that a counter-suit in the Privy Council was impossible as the verdict was "merely an expression

---

[93] Ibid., p. 171.
[94] See Karl Marx, *Capital*, Vol. I, trans. Ben Fowkes (New York, NY: Vintage, rpt. 1977), p. 179.
[95] Vide, *ILR*, Calcutta Series, Vol. XLVIII, January to December 1921, p. 173.
[96] See *The Indian Planters' Gazette and Sporting News*, June 5, 1920, p. 754, NAL.
[97] Ibid., p. 754.

of opinion given for the guidance of the Revenue Authorities."[98] It was neither an order nor a decree. Amidst the hullaballoo regarding apportioning taxable income, the GOI deputed a committee comprising Mr. H. E. Spry, ICS of the GOB, Mr. G. G. Sim, ICS of the Finance department, Mr. J. A. Milligan of the Assam Labor Board, and Mr. T. C. Crawford of the ITA to report on the impasse.[99] Their recommendations and a subsequent compromise between the government and the tea association saw some results – it was agreed under the Indian Income Tax (Revised) Act XI of 1922 that 25 percent of the income from production would be treated as business income and therefore liable to the impost.[100] It is not entirely clear how this figure was arrived at. Of course, the ITA protested that the 25/75 arrangement was decidedly illegal. Unlike other cash crops, say cotton for instance, tea, the association claimed, could not be kept for an indefinite duration and then be made fit for use. In other words: "the cultivator of tea is and must be the manufacturer."[101] The government acquiesced and dropped these percentage figures. In its place, "a more general rule" was instituted whereby assessing tax officers were left to decide what, in their opinion, constituted a "reasonable profit" on these respective operations.[102]

To be sure, the dust never really settled on the matter between the government and the planters (and the ITA). Beyond apportioning agricultural and industrial incomes, disputes intermittently arose regarding the tax status of employee benefits (rent-free accommodation, for instance), and other fixed and moveable assets of tea concerns. Separately, Mr. W. G. McKercher, General Manager of the Amgoorie tea estates in Assam and Vice Chairman of the Assam Branch of the ITA in his deposition to the Royal Commission on Agriculture in India (hereafter RCAI) in 1927 argued that excessive taxation on the tea enterprise – that included export duties, income tax, duty on petrol, and import duty on tea boxes among others – had stalled development

---

[98] Vide, letter no. 10300, Calcutta, June 28, 1920 from M/S. Orr, Dignam & Co., to ITA in *Detailed Report of the General Committee of the Indian Tea Association for the Year 1920* (Calcutta: Criterion Printing Works, 1921), pp. 138–139, Assam State Archives (hereafter ASA), Guwahati, Assam.

[99] Vide, Circular No. 84, Calcutta, December 7, 1921, from ITA to All Members of the Association in *Detailed Report of the General Committee of the Indian Tea Association for the Year 1921* (Calcutta: Criterion Printing Works, 1922), p. 222, ASA.

[100] See Sir Percival Griffiths, *The History of the Indian Tea Industry*, p. 559.

[101] See *The Indian Planters' Gazette and Sporting News*, June 5, 1920, p. 754, and *The Indian Planters' Gazette and Sporting News*, June 12, 1920, p. 782, NAL.

[102] Quoted in Draft Paper of the Finance Department, GOI, Lord Winterton to Mr. David Grenfell, MP, No. F/6297/27 dated October 5, 1927, MSS EUR F 174/1217, Asian and African Studies, British Library, London.

Agriculture or Manufacture? 73

works in Assam.[103] Calling themselves the "black sheep" of the province, McKercher suggested that the export duty, amounting to 30 lakh rupees in 1927, was especially ruinous, and should have been withdrawn after its intended role as war measure had been served. For him, central taxation on the tea enterprise was counter-productive for local progress on all fronts, agriculture included. In his assessment: "we pay treble the revenue . . . still we are useless to the Province."[104] If the latter was a view often shared by the Assamese, McKercher was convinced that the absurd logic of the income tax only added injury to the wound.[105]

More to the point, planters and the ITA protested that under prevailing stipulations, too much power and authority had been divested to the tax officers to fix the ostensibly indivisible – the agrarian and manufacturing portions of tea profits. Eager to draw curtains on this long-standing confusion and conundrum, the GOI realized in 1927 that only a clear ruling on tax percentages would close this debate. Put differently, the law had to be amended to give the Central Board of Revenue "definite power to prescribe how much of income derived in part from agriculture and in part from business."[106] The amending bill received the Governor-General's assent on September 22, 1927.[107] Accordingly, it was now fixed that:

Income derived from the sale of tea grown and manufactured by the seller shall be computed as if it were income derived from business, and 40 per cent of such income shall be deemed to be income, profits, and gains liable to tax.[108]

---

[103] See statements of Mr. W. G. McKercher, *Royal Commission on Agriculture in India: Evidence taken in Assam*, Vol. V (London: His Majesty's Stationery Office, 1927), pp. 205–228.
[104] Ibid., p. 217.
[105] McKercher remarked on the income tax apportionment: "you may say that we have a factory, simply because we have got a steam-engine or something, and just get more money out of us," in *Royal Commission on Agriculture in India: Evidence taken in Assam*, 216; according to McKercher's statistics, the tea industry contributed 30 lakh rupees in export duty, 15 lakh rupees in income tax, and 10 lakh rupees in import duty (on tea boxes) to the central government in 1926–7, in ibid., p. 209.
[106] Quoted in letter marked "Confidential," No. 666-I.T./26, from Deputy Secretary to the GOI to Her Majesty's Under Secretary of State for India, Financial Department, India Office, London, dated August 24, 1927, p. 2, in MSS EUR F 174/1217, Asian and African Studies, British Library, London.
[107] This became Act XXVIII of 1927; effectively, only section 59 of the 1922 Act was amended to empower the Central Board of Revenue to decide on these figures, see *A Collection of the Acts of the Indian Legislature for the Year 1927* (Calcutta: Govt. of India Central Publication Branch, 1928).
[108] Quoted in *The Gazette of India*, November 19, 1927, Part I, p. 1057; the percentage figure had already been decided by September 30 of that year, in MSS EUR F 174/1217, Asian and African Studies, British Library, London.

This apportionment remained in place till after India's independence in 1947.[109]

Tea was a political plant, and was caught up in high intrigue throughout the income tax debates.[110] *Camellia sinensis* was at the center of the dispute. For planters and the ITA, the tea enterprise was not a sum of its parts; the parts were the sum. In other words, they consistently claimed that the essential characteristics of the tea plant did not mutate during the mechanical phases of production. Its inherent quality – as leaf – was untouched and the marketed substance only appeared transformed in shape, color, and texture. Botanically speaking, the plant and the plantations were, for planters at least, an agricultural setup. For sure, the

---

[109] Though outside the purview of the present study, it is important to consider the larger fiscal context of the income tax debates. Economic historians of India have argued that, prior to World War I, the bulk of the government taxes accrued from land revenue (55.5 percent of total income in 1855–6). More "modern" taxes, such as the income tax, customs, and excise contributed very minimally (12 percent of total income in 1858–9) to public finances in British India. One crippling effect of this heavy reliance on land revenue, Tirthankar Roy suggests, was that the government budget remained inflexible and inelastic. The two other commodity taxes – on opium and salt – were also income inelastic. In other words, even as the economy expanded, corresponding revenue from these sources did not match up. As a result, in comparison to tropical British and French colonies in Asia, the colonial government's revenue as a proportion of the GDP was extremely low. Roy argues that this changed around 1919 when the importance, and contribution, of the land tax declined markedly (fall in the value of agricultural produce, sustained campaigns by landed merchants and associations, inflexible incomes from permanently settled areas, and political considerations in the *Ryotwari* areas aiding in this respect). In its place, income tax, customs and excise took on renewed significance as far as the GOI's finances were concerned:

	1858–9	1920–30 (average annual)
Percent of Total Revenue		
Customs	8	26
Land Revenue	50	20
Salt and Opium	24	neg.
Excise	4	17
Income Tax	0.3	10

Quoted in Tirthankar Roy, *The Economic History of India, 1857–1947*, third edition (New Delhi: Oxford University Press, 2011), pp. 252–257; see also, Dharma Kumar, "The Fiscal System," in Dharma Kumar and Meghnad Desai, eds. *The Cambridge Economic History of India, Vol. 2, c. 1751–1970* (Cambridge: Cambridge University Press, 1983), pp. 905–944; economically, the GOI's insistence on the tea income tax has to be read in light of this historical context.

[110] I borrow the term "political plant" from Londa Schiebinger's study of *Poinciana pulcherrima*, commonly known as the "peacock flower" in Schiebinger, *Plants and Empire: Colonial Bioprospecting in the Atlantic World* (Cambridge, MA and London: Harvard University Press, 2004).

commercial success of tea depended on flavor – an attribute of its own qualities as leaf and not of the finished product of its seed or fruit.

As the first section of this chapter shows, the advancement of "technical" and "mechanical" know-how in the tea enterprise could not dispense with its botanical element.[111] Of course, separating tea's cultivated and industrial remnants was a tautological and imprecise exercise, practically and otherwise. Given this long-drawn historical wisdom, and especially the government's early "blunder" with the China plant, the planting community (including the ITA) claimed *natural* authority over all aspects of the industry – product, character, and operation.[112] As Edgar Thompson once remarked: "the plantation arises as the personal 'possession' of the planter, and it is from the standpoint of his interests that the course of its development is directed. It is first of all a unit of authority."[113] As the enterprise's self-imputed spokesmen, the GOI's contrarian views and fiscal posture on tax apportionment therefore appeared threatening at many levels. It had ruptured this zealously guarded, and so-called experts-sanctioned jurisdiction over the commodity and its production space.

But not all loose ends of this argument had been tied up for the anti-government faction. As briefly indicated earlier, the 1886 and 1918 tax acts precipitated a theoretical double-bind for planters and the ITA. This too was linked with the validation, indeed exhortation of the indigenous plant's alleged supremacy over its Chinese variant. As Indian tea found acceptance – among planters, the metropolitan palette and the market – after 1870, an elaborate ideology of refinement was concurrently drawn up. In addition to racial ("English superintendence") and climatic causes mentioned above, automation was championed as leading to its alleged purity and "cleanliness." Lieutenant Money thus argued in 1884: "the Tea of Hindustan is now all manufactured by machinery, but in China it is hand-made."[114] In other words, paradoxically, "not being touched by hand at all" during the final stages was extolled in his idea of *natural* superiority. As far as Money was concerned, this was Indian tea's decided advantage over the "dirty" and "hard muscular exertion" still practiced by the Chinese.[115] In the world of Indian tea, machinery were not benign

---

[111] On this point, I disagree with Jayeeta Sharma's argument to the contrary; see Sharma, "British Science, Chinese Skill and Assam Tea: Making Empire's Garden," p. 443.
[112] On an interesting debate regarding the "natural" in indigo in colonial India, see Prakash Kumar, "Plantation Indigo and Synthetic Indigo: Redefinition of a Colonial Commodity," *Comparative Studies in Society and History* 58, 2 (April 2016): 407–431.
[113] See Edgar T. Thompson, "The Climatic Theory of the Plantation," *Agricultural History*, Vol. 15, No. 1 (January 1941): 49–60.
[114] Lieutenant-Colonel Edward Money, *The Tea Controversy*, p. 9.   [115] Ibid.

objects; they were steeped in ideology, propaganda, and, in our case, irony.[116]

The income tax turf wars were untenable at all levels, and on all fronts. As a semantic and operational "return to nature" became expedient for planters and the ITA during 1886 and 1918, the history of the enterprise – and the tea plant – showed that neither "agricultural" nor "industrial" summed up its internal distinctions and nuances. The tea experts consciously or unconsciously invoked by planters in defense were themselves ambivalent on this question. Sometimes, these "authorities" of tea planting spoke in a forked tongue, as Lieutenant-Colonel Money's arguments demonstrate. More importantly, and embarrassingly, denying mechanization's novelty – indeed modernity – in the tea-manufacturing process contravened and contradicted the planting community's own proclamations of advancement and progress.

On the government's part, apportioning the "first part" of the enterprise as agrarian and the "latter part" as industrial was convenient in theory and law but complicated, even incommensurable in practice. Assigning exact numbers to these processes (whether 25/75 in 1922 or 40/60 in 1927) was even more problematic. Sir Mookerjee, in invoking Crole and Money to make his case, was scrupulous in his sources but selective in his interpretation.

Relegated as a footnote in Indian tea's economic history, the income tax debates were much more than a fiscal argument. The commercial, botanical, ideological, and political aspects of this conundrum – though connected – did not always converge. The deliberations of 1886, 1918 and 1920, read alongside the vast literature on tea cultivation and manufacture provide a fascinating glimpse of this contested, but forgotten, relationship between the plant, the planter, and the plantation in colonial India.

---

[116] To be sure, tea publicity went beyond simple market rivalry. For colonial administrators, the ITA, and planters, the enterprise had bestowed Assam and British India its gift of "progress" and modernity – horticultural, social, and material. The introduction of mechanization was a necessary, indeed celebratory step in this march of "improvement." For instance, Sir Edward Gait, Assam's census commissioner and noted civil servant argued: "the benefits which the tea industry has conferred on the Province have been many and great. The land most suitable for tea is not adapted to the cultivation of rice, and the greater part of it would still be hidden in dense jungles if it had not been cleared by the tea planters ... the literate classes have obtained numerous clerical and medical appointments in the gardens ... a great impetus has also been given to trade, and new markets have been opened in all parts of the Province. The existence of the tea industry has been a potent factor in the improvement of communication by rail, river and road," in Gait, *A History of Assam* (Calcutta: Thacker, Spink, rpt. 1967), p. 413. As mentioned, claiming indivisible agrarian status for the tea enterprise belied, and contradicted, these other ramifications of tea's supposed modernity.

3 Bugs in the Garden

Wherever humans have broken ground, whatever frontiers humans have explored, they have discovered that they are latecomers, following in the six-legged footsteps of insects[1]

If the fiscal debate on the "natural" presents a macro view of the tea industry's agro-economic problems, it had other minute, and largely unseen, competitors to deal with. For sure, tea pests and blights appeared almost concurrently with the start of the Assam plantations. C. A. Bruce, acknowledged pioneer of tea planting and manufacture,[2] remarked on the mole cricket in his famous 1838 account of the Singpho and Muttock tea tracts of upper Assam. Experimenting with tea seeds and young saplings in the hot summer sun, Bruce noticed the insect's depredations in nipping off the tender leaves and depositing them underground near its root.[3] The tea plant's prospects were observably bleak.

The tea mosquito bug (*Helopeltis theivora*), the red spider (or tea mite, *Tetranychus bioculatus*), thrips (*Scirtothrips dorsalis*), tea aphis, looper caterpillars, and blister blight particularly troubled Assam planters in the period under study. Though not an exhaustive list, these were some of the major predators of the tea plant and caused the maximum damage. Broadly, the focus on tea pests and climate in this chapter unearths the largely forgotten relationship between this plantation economy and its built environment. Literal and proverbial bugs[4] in the garden, they

---

[1] May Berenbaum, *Bugs in the System: Insects and Their Impact on Human Affairs* (Reading, MA: Helix Books, 1995), Preface, p. xii.
[2] Antrobus, *A History of the Assam Company*, p. 22.
[3] C. A. Bruce, *An Account of the Manufacture of the Black Tea as now Practiced at Suddeya in Upper Assam, By the Chinamen Sent Thither for that Purpose, with Some Observations on the Culture of the Plant in China and its Growth in Assam* (Calcutta: Bengal Military Orphan Press, 1838), p. 15.
[4] The words "bug" and "insect" might now seem interchangeable but this has historically not been the case. May R. Berenbaum writes: "Approximately four centuries after Aristotle's *Historia Animalium*, Pliny the Elder offered his interpretation of insect classification in the form of his magnum opus *Historia Naturalis* . . . despite its inaccuracies, [this]

78 Tea Environments and Plantation Culture

Figure 3.1 *Helopeltis theivora* Waterhouse (Tea mosquito) © Rison Thumboor, Wikimedia Commons, CC BY-SA 4.0

gnawed away at profits, ideas of scientific control, and expert proclamations of agrarian improvement. Indeed, these tea blights were not externally induced but grew out of the natural conditions of the Assam estates (especially shade-trees and rains) and eventually dispersed through the body of laborers, winds, plucking baskets, birds and the very structures that supported the industry. In other words, these were not bugs that

was the authoritative source on natural history for the next 1,400 years. Medieval compilations borrowed heavily from his text and few innovations were made during the Middle Ages. For example, Bartholomaeus Anglicus (name notwithstanding, a Frenchman) compiled nineteen volumes around AD 1230 entitled *De Proprietatibus Rerum*. The work was intended to be a complete description of the universe ... the word 'bug' dates back to this era and refers to a ghost or hobgoblin – something difficult to see and vaguely unpleasant (a term quite apt for most insects medieval people were likely to encounter). The word 'insect,' on the other hand, entered the English language only in 1601, when Philemon Holland published a translation of Pliny's *Historia Naturalis*. A year later, Ulysses Aldrovandus, an Italian, introduced a few taxonomic innovations of his own. Insects were divided according to habitat into Terrestria, or land-dwelling species, and Aquatica, the water-dwelling species. Each group was further divided according to the presence or absence of appendages (Pedata and Apoda, accordingly) and then subdivided further according to whether wings were present or absent (Alata and Aptera, respectively) ... In 1758, Carl Linné published a book called *Systema Naturæ*, in which he used a binomial, or two-name, system, consistently for the first time. The system so impressed people that it was universally adopted; no scientific names published before Linnaeus's time are considered valid and all subsequent names have conformed (or must continue to conform) to the Linnaean system," in Berenbaum, *Bugs in the System*, pp. 3–4.

came from afar.⁵ As crop quality, and ultimately profits, dwindled in the face of these unknown – and little understood – ecological competitors, planters in Assam sought out local methods of pest management and eradication, forged alliances with pan-imperial peers on questions of remedy and balked at pedagogic, but mostly ineffectual laboratory "fixes" and metropolitan know-how on the subject. These interactions also allow us to rethink the relationship between human actors and non-human participants in the making of one of British Empire's most coveted objects of desire.⁶ If high incidence of malaria, cholera and kala-azar (or black fever) deaths among tea laborers show the social ramifications of agronomic manipulation, the continual resurgence of tea pests and blights demonstrate that profits and pathogenicity were centrally linked in this commodity story.⁷

This history of plant (and later human) pathogens lends itself to the critical sociology developed by John Law, Bruno Latour and others around the late 1980s. Better known as the "Actor–Network Theory" (ANT), it argued for a radical reconfiguration of social agency – and our understanding of that agency – as *products* of innumerable "networks," "actors," and "assemblages" that included animals, machines and non-humans.⁸ In their view, social structures, organizations and processes were never constituted one-directionally by autonomous, all-knowing and wholly-in-control actants. For Latour, social reality and institutions (including capitalism) were part of, and resulted from a series of "associations" with no one designated center or locus of authority and power.

---

⁵ See William Cronon, "The Uses of Environmental History," *Environmental History Review*, 17(3) (1993): 10.
⁶ See Richard White, "Discovering Nature in North America," *Journal of American History* 79 (1992): 874–891; also, White, *The Organic Machine: The Remaking of the Columbia River* (New York, NY: Hill and Wang, 2005); Matthew Mulcahy, *Hurricanes and Society in the British Greater Caribbean, 1624–1783* (Baltimore, MD: The Johns Hopkins University Press, 2006); J. R. McNeill, *Mosquito Empires: Ecology and War in the Greater Caribbean, 1620–1914* (Cambridge: Cambridge University Press, 2010); Robert E. Kohler, *Lords of the Fly: Drosophila Genetics and the Experimental Life* (Chicago, IL: University of Chicago Press, 1994); Timothy Mitchell, "Can the Mosquito Speak?" in *The Rule of Experts: Egypt, Techno-Politics, Modernity* (Berkeley, CA: University of California Press, 2002); Paul S. Sutter, "Nature's Agents or Agents of Empire? Entomological Workers and Environmental Change during the construction of the Panama Canal," *Isis* 98(4) (2007): 724–754 as further case-studies of this human-non-human overlap. See William Beinart and Lotte Hughes, *Environment and Empire* (Oxford: Oxford University Press, 2007), p. 10 for a discussion of commodity "desires" during the high imperial period.
⁷ I discuss labor health and disease environments in the following chapter.
⁸ See Bruno Latour, *Reassembling the Social: An Introduction to Actor-Network Theory* (New York, NY: Oxford University Press, 2005); also Latour, *We Have Never Been Modern* (Cambridge, MA: Harvard University Press, 1993). See also John Law and J. Hassard, eds., *Actor Network Theory and After* (Oxford: Blackwell, 1999).

ANT is especially helpful in unpacking the asymmetries of economic activity. In the case of bugs, for instance, our discussions below show that the political economic structure of the Assam plantations was a web of alliances, actors and forces. While some of those relations – between labor and capital, for example – could be controlled and manipulated with some degree of certainty, others – between pest infestation and tea microclimate – were difficult, indeed impossible, to separate and predict. As it is, the exponential rise of blights in the Assam gardens brought home the uncanny realization that planters and tea "experts" were neither fully in the know nor unchallenged masters of this commodity enterprise. The role of formal entomological "science" in managing bugs is even more suspect. For not only was it unable to provide sustainable remedy in most cases, technical fixes such as chemical pesticides only *created* problems that did not exist before. Finally, these mediations between the human and the non-human allow us to detect that the costs and consequences of *this* plantation economy were not merely physical and human, but ecological and economic at the same time.

Methodologically, of course, ANT has several limitations.[9] While the entire gamut of those criticisms need not be revisited, some assessments are in place. One of the most important limitations of ANT is that is does not allow for situating responsibility and accountability in social actions.[10] As I show below, the bugs in the Assam gardens were not free-floating, a-historical arrivals in the plantation enterprise, but arose *as a result of* the structures of the industry. To that end, the profit mandate of these imperial estates, the scramble for and unthinking take-over of lands, large-scale irrigation works, unhealthy forest cover, and monoculture cash cropping were directly or indirectly responsible for pest occurrence and devastation in the region. Secondly, ANT's analysis of power as endlessly "relational" precludes our ability to historicize the origins and effects of that power.[11] In other words, these bugs were not just part of the human/non-human networks of tea economics, but *productive* constraints and consequences of the Edenic ideology of "improvement." If political economy is a network of relations, some actor–networks are more equal

---

[9] See Donna J. Haraway, *Modest_Witness@Second_Millenium.FemaleMan©_Meets_OncoMouse*™ (London: Routledge, 1997); Steve Fuller, "Why science studies has never been critical of science: Some recent lessons on how to be a helpful nuisance and a harmless radical," *Philosophy of the Social Sciences* Vol. 30, No. 1: 5–32; and Scott Kirsch and Don Mitchell, "The Nature of Things: Dead Labor, Non-Human Actors, and the Persistence of Marxism," *Antipode* Vol. 36 (2002): 687–705 to name a few.

[10] See Kirsch and Mitchell, "The Nature of Things," p. 689.

[11] As Kirsch and Mitchell argue, this attitude, in principle, forecloses the search for, and the impact of those effects of power that lie beyond the horizons of its human and non-human actors, in ibid., p. 692.

and more powerful than others.¹² Suggesting otherwise flattens the discourse of power, especially colonial power and its impact on human lives and biosocial engineering in the Empire.

Bugs in the Garden

Samuel E. Peal, a planter in the Sibsagar district was perhaps the first to draw attention to the tea bug, an arthropod that resembled the common mosquito.¹³ He presciently cautioned that this pest was to be the tea planter's greatest enemy in the years to come and had the potential to seriously cripple the industry and reduce yield. The warning was clear: "those who [were] indulging in dreams of thirty and forty percent [would] soon be roused up when they [found] their profits reduced to three or four."¹⁴ With seven accompanying color plates in the *Journal of the Agricultural and Horticultural Society of India* (hereafter JAHS), Peal recorded his observation of the pest's physiognomy and impact on tea leaves and shoots. What worried him more was the bug's eco-biology, a vicious parasitism that allowed it to grow and draw sustenance from the tea plantation habitat. He thus debunked the theory that excessive shade or lack of jungle clearing led to an increase in the tea mosquito pest. Drawing on infestation case studies from gardens that were relatively open, and from those recently cleared, Peal provided the damning conclusion that the very conditions necessary for successful tea harvests created the host environment for the bug.¹⁵ While Peal was in no position to offer scientific remedy, he astutely recommended against adding labor-hands for physical removal of the pest or syringing tea leaves with

---

[12] Ibid., p. 690.
[13] S. E. Peal, "The Tea Bug of Assam," *Journal of the Agricultural and Horticultural Society of India* (New Series) Vol. 4, No. 1 (1873): 126–132.
[14] Ibid., p. 126. Samuel Peal is also reported to have written on the blister blight of tea as far back as 1868 but this source remains untraced; quoted in Harold H. Mann, "The Blister Blight of Tea," Indian Tea Association Circulars, No. 3 (Calcutta, 1906), 1; MSS EUR/ F 174/11, Asian and African Studies, British Library, London. It is also important to note here that entomology and empire are closely connected. In fact, J. F. M. Clark argues: "Economic entomology achieved professional respectability between 1880 and 1914 through the creation of specialist educational programmes and acknowledged posts in the field. The identification of insects as vectors of disease – the emergence of medical entomology within the rubric of tropical medicine – provided a further strong rationale for the study of applied entomology. Experience of insect control and eradication in empire shaped the careers, knowledge and practices of British entomologists. As an institution or discipline, applied entomology in Britain was forged from agricultural science and tropical medicine, under the umbrella term of economic entomology," in J. F. M. Clark, *Bugs and the Victorians* (New Haven, CT and London: Yale University Press, 2009), p. 188.
[15] S. E. Peal, "The Tea Bug of Assam," p. 128.

medicinal decoctions. The futility of these measures was not lost on Peal, for Assam's torrential monsoonal rains regularly washed away these fluids and created the perfect moisture-base needed for the tea bug's increase. With resigned hope, he wrote: "I see no cure till Nature produces her own, in good time; and one is certain to come in the end, though probably not under twenty to fifty years."[16]

The mutually conducive (and occasionally harmful) ecosystem for tea growth and pest development remains a complex and historically interesting environmental backdrop to the Assam plantation story. Size and capacity for damage were often inversely proportional. In the case of the tea aphis, for instance, planters often wondered how an insect barely observable by the naked eye could propagate with such rapidity and inflict widespread destruction at the same time.[17] The question of agency was pressing, and James Wood-Mason, deputy superintendent of the Indian

---

[16] Ibid., p. 130; admittedly, Peal was an exceptional figure in the pantheon of early Assam planters. As naturalist, ethnographer, ornithologist, and geographer, Peal distinguished himself in an occupation otherwise much debased in nationalist, metropolitan and elite British *imaginaire* as that given to the pleasures of the body and mind. It is interesting, for instance, to counterpose figures like Peal with Maurice Hanley, Charles Webb or the fictitious Beth and McLean planter sahibs of *Kuli Kahini* and *Cha-kar Darpan* respectively; see Maurice Hanley, *Tales and Songs from An Assam Tea Garden* (Calcutta and Simla: Thacker, Spink and Co., 1928); Ramkumar Vidyaratna, *Kuli Kahini*, ed. Biswanath Mukhopadhyay (Calcutta: Jogomaya Publications, 1886); and Dakshinacharan Chattopadhyay. *Cha-kar Darpan Natak* in *Bangla Natya Sankalan* (Calcutta, 2001), for a discussion of these other characters. Made a Fellow of the Royal Geographical Society, Peal was eulogized as "An Assam Naturalist" in his obituary of August 12, 1897. The contributor records: "it was perhaps a mistake that Mr. Peal was a tea-planter at all. He was essentially a naturalist," in Obituary, *The Journal of the Polynesian Society* Vol. 6, No. 4 (1897): 216–218; reprinted from *Calcutta Englishman*, August 12, 1897; for a fascinating extension of this point, see the paper by Tony Ballantyne, "Mr. Peal's Archive: Mobility and Exchange in Histories of Empire," in Antoinette Burton, ed., *Archive Stories: Facts, Fictions and the Writing of History* (Durham, NC: Duke University Press, 2005), pp. 87–111. And these were no empty signifiers or unthinking sobriquets either; Peal was a regular contributor to the *JAHS*, the *Proceedings of the Royal Geographical Society, Science, Nature*, the *Journal of the Royal Asiatic Society* and many others. Journal contributions aside, it is not surprising that Peal is also credited as being the discoverer of the Peal Palmfly or *Elymnias peali* classified by Wood Mason in 1883, cited in "Description of a new Species of the Lepidopterous Genus Elymnias," J. Wood-Mason quoted by Major G. F. L. Marshall and Lionel De Nicéville, *The Butterflies of India, Burmah and Ceylon* (Calcutta: The Calcutta Central Press, 1882), p. 277 and is even reported to have provided information on rich deposits of coal and petroleum in the Margherita region of upper Assam, cited in Rajen Saikia, *Social and Economic History of Assam 1853–1921* (New Delhi: Manohar, 2000), p. 151. In a way, Peal was a planter only by default. His occupational residence in Sibsagar afforded a rich and seemingly inexhaustible ecological laboratory that connected him to the world of tea, science, ethnography, and entomology all at once.

[17] Reprinted in section on "Tea Blights and Pests" in *The Tea Cyclopaedia: Chapters on Tea, Tea Science, Blights, Soils and Manures, Cultivation, Buildings, Manufacture Etc., With Tea Statistics*, (London: W. B. Whittingham & Co., 1882), pp. 34j–66.

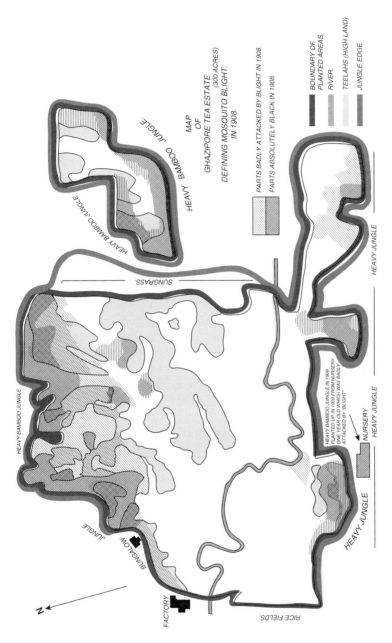

Map 3.1 Map showing tea mosquito bug attack on Ghazipore tea estate, 1908. The dark shaded portions show areas affected, with the darkest spots indicating severe damage © Adapted from C. B. Antram, *The Mosquito Blight of Tea: Investigations during 1908–09*, Indian Tea Association (Calcutta: The Catholic Orphan Press, 1910)

Museum, Calcutta suggested inter-insect dispersion as partly responsible for pest occurrence in the Assam estates.[18] However, climate and agro-ecology were not always beneficial allies to these pests and could turn against each other depending on circumstance. Small tea pests like the aphis were regularly, though not always, washed away or killed by heavy downpours or due to periods of prolonged drought and dryness. More than two thousand miles to the south, a strikingly similar problem befell coffee planters in Ceylon (now Sri Lanka). As this British plantation economy took over the island's middle highlands, several ecological challenges in the form of parasitic fauna came into focus. The coffee rat (*Golundus ellioti*) wreaked havoc, as did coffee rust, and the black, brown, and white bug. As in Assam, natural allies both arrested and spread bug infestation in the Sri Lankan plantations.[19] Extensive burning, or "grubbing them out" with a mamoty produced attendant problems of soil erosion and loss of nutrients.

Surprisingly, the depredations of the tea mosquito bug in Assam caught the attention of the Calcutta scientific establishment only after a decade of Peal's 1873 article. Wood-Mason was instructed to carry out a detailed field study and his report was finally submitted on June 8, 1881.[20] While repeating some of Peal's observations verbatim, Wood-Mason's study was based more on laboratory cross-examination of facts. He suggested a vigorous and unremitting removal of blighted portions of the tea plant, a move that required adding to the already demanding labor working hours of these estates. He also advanced the hypothesis that the olfactory quality of tea juice provided differential immunity from the mosquito bug. The rasping and pungent liquor of the native Assam plant allegedly rendered it immune from attack while the milder extract of the Chinese variant made it more susceptible to damage.[21] These ideas were, however, to be vigorously disproved by successive waves of the tea bug assault on all species of tea in Assam. In hindsight, Wood-Mason's report remained rather inconclusive and haphazard, though it did provide some interesting insights and analysis of the tea mosquito bug. More importantly, his report was one of the earliest to flag the tea mite (commonly known as the red spider) as a parasite of the Assam tea leaf.

---

[18] Ibid., p. 38.
[19] See James L. A. Webb, Jr., *Tropical Pioneers: Human Agency and Ecological Change in the Highlands of Sri Lanka, 1800–1900* (New Delhi: Oxford University Press, 2002).
[20] James Wood-Mason, *Report on the Tea-Mite and the Tea-Bug of Assam* (London: Taylor and Francis), 1884.
[21] Ibid., p. 18.

The effects of the red spider on tea growth were reported to be far more devastating.[22] Wood-Mason observed that the mite lived in small "societies" on the upper surface of full-grown leaves, and beneath a delicate web that it spun for itself as protection. Providing shelter and survival from the heavy April rains, this skein allowed the spider to go unchecked and unnoticed. While the intriguing relationship between rains and remedy in the Assam gardens have already been commented upon, it was more amply evident in the case of the tea mite. For instance, long periods of torrential showers often broke up the intricate web and led to periods of pest disappearance. But this was hardly a workable curative strategy. Wood-Mason's report authoritatively demonstrated that the red spider, although of genus Capsidae characteristic of Indo-Malayan fauna, was not an alien import but an indigene of the Assam tea country.[23] This view also confirmed Peal's initial suspicion of the mutually beneficial host conditions of the tea plant and pest in the Assam gardens.[24] Peal reiterated in *The Indian Tea Gazette* that the red spider was one of the oldest, most universal and widely distributed pests in operation – ranging from the sea level to snow-capped mountain ranges of the upper Himalayas.[25] A later study on the bionomics of the red spider showed that the mite continued to breed during the cold season and could be found at all stages of the tea plant growth.[26] This made it clear that among the factors influencing the incidence of red spider and the intensity of attack, weather conditions were preeminent.[27] The more insidious aspect the mite was the manner

---

[22] For a scientific study on the red spider and its relationship to the tea plant, see G. M. Das, "Bionomics of the Tea Red Spider, *Oligonychus Coffeae* (Nietner)," *Bulletin of Entomology*, Vol. 50, No. 2 (1959): 265–274.

[23] Wood-Mason, *Report on the Tea-Mite and the Tea-Bug of Assam*, p. 13.

[24] A recent scientific study reiterates this by suggesting a further correlation between age, acreage and pests. It demonstrates that the microclimate of the monoculture tea crop provides a continuous food source for various kinds of "phytophagous arthropods," reaching a saturation level after 35 years of growth. Statistically, the findings show that northeast India harbors the largest number of tea pest species (250) which directly corresponds to area (361,663 acres in 1981) and tea age (138 years). The research suggests that most tea pests are recruited "locally," with only about 3 percent being common across regions. See Barundeb Banerjee, "An Analysis of the Effects of Latitude, Age and Area on the Number of Arthropod Pest Species of Tea," *Journal of Applied Ecology* Vol. 18 (1981): 339–342.

[25] Reprinted in section on "Tea Blights and Pests" in *The Tea Cyclopaedia: Chapters on Tea, Tea Science, Blights, Soils and Manures, Cultivation, Buildings, Manufacture Etc., With Tea Statistics* (London, 1882), p. 38.

[26] Das, "Bionomics of the Tea Red Spider, *Oligonychus Coffeae (Nietner)*," *Bulletin of Entomology*, Vol. 50, No. 2 (1959): 265–274.

[27] Ibid., p. 272.

of its dispersion within the tea estates: wind, cattle, goats, birds and other insects[28] being among the chief agents of circulation. Even laborers working on the plantations were indirectly responsible as the red spider spread unnoticed through clothing and tea baskets.[29]

At the turn of the century, blister blight proved to be a severe and crippling concern for planters in Assam. A fungal disease, it struck with particular virulence in April and May 1906. Dr. Harold H. Mann, scientific officer to the ITA published a report on the blight that year after his visits to the affected upper Assam districts. He noted that the impact of the fungus was localized in scope but epidemic in character. Commenting on this peculiarity, Mann observed that the climatic and soil conditions of the districts under siege (namely North Lakhimpur, Golaghat and Jorhat) were directly responsible for the intensity of infection.[30] The relative immunity of the other tea districts from the blister virus that year only made clear the challenges of adopting a region-wide approach to pest reduction and control. Interestingly, W. McRae, mycologist to the Government of Madras, commissioned to study the outbreak of blister blight in the neighboring Darjeeling district in 1908–09, argued that the fungus was "new" to the tea region despite being "detected" and "confined" to the Brahmaputra valley as early as 1895.[31] Adding to our knowledge of the restricted nature of the disease, McRae observed that the extent of damage was often dependent on the tea variety (or *jat*) – the high-quality Assam and hybrids being the most susceptible and the Chinese and Manipuri variants relatively immune. McRae reiterates and confirms Mann's earlier hypothesis of the relationship between rainfall, pruning and blister attack: "the greater loss is attributable to wet, unfavourable weather in July and August ... the worst damaged piece of tea was a heavy pruned block."[32] He also suggested provocatively that, while the exact cause of the fungus in Darjeeling was not definitively known, it might have been "imported" from Assam valley by tea-seed transfer among other ecological and human factors.[33]

---

[28] Wood-Mason, however, disagreed on this widely held notion of inter-insect agency by planters. He claimed, somewhat emphatically, in his report that "mites do not commonly occur parasitically on the outside of the bodies of the most diverse group of insects," in Wood-Mason, *Report*, p. 10.

[29] Das, "Bionomics of the Tea Red Spider," p. 272.

[30] Harold H. Mann, *The Blister Blight of Tea*, Indian Tea Association Circular No. 3/1906, MSS EUR/F/174/11, Asian and African Studies, British Library, London.

[31] W. McRae, "The Outbreak of Blister-Blight on Tea in the Darjeeling District in 1908–1909," ITA Circular No. 3/1910, MSS EUR/F/174/1517, Asian and African Studies, British Library, London; interestingly, there is no mention about the 1868 chapter on the blister blight by S. E. Peal in McRae.

[32] Ibid., p. 6.   [33] Ibid., p. 7.

In addition to the above, thrips also damaged tea in Assam and neighboring districts during this period.[34] Reproducing exponentially in the shade of the tea bush, thrips arrested the growth of young leaves and shoots. The more worrisome feature of the insect was that it hardened the leaf and made it brittle, thereby leading to a recognizable reduction and "loss in flavour."[35] For a commodity that relied on taste as its distinctive hallmark, this was a serious discovery.

Beyond entomological findings and planter reports, the proverbial bug in the empire's garden found its way into government correspondences, revenue proceedings and annual tea balance sheets. While many factors, including political climate, seed quality, methods of plucking, labor, mortality and machinery, contributed to fluctuations in tea production, the trio of pests, rainfall, and climate impacted relentlessly in terms of both quality and volume. Interestingly, reporting on the ravages of hailstorms and red spider blights in 1883, C. J. Lyall, then officiating secretary to Assam's chief commissioner, critiques James Wood-Mason's pest experiments as esoteric laboratory science far removed from the practical and pragmatic challenges to planters on the ground.[36] The situation spoke for itself. Consider the following figures in Table 3.1 below for changes in tea yields during a ten-year period (1884–95) in some of the most important tea producing districts of Assam.

To be sure, the Assam tea enterprise was a vast and complex operation, and no one component influenced variations in production and total output.[37] Amalgamation of smaller estates into bigger holdings, finer plucking, rise in labor expertise, use of machinery, demand, and overharvesting, among others, significantly altered numbers in terms of acreage and outturn. Three factors, however, remained consistently important in causing these fluctuations, namely rainfall, pests, and weather conditions. For instance, unpredictable monsoons, prolonged drought and mosquito blights in 1884 severely reduced the yield in Nowgong and Cachar, while dry weather and selective plucking in Lakhimpur around 1887 changed

---

[34] C. B. Antram, "The 'Thrips' Insects of Tea in Darjeeling: Investigations During the Season 1908," ITA Circular No. 3/1909, MSS EUR/F/174/1516, Asian and African Studies, British Library, London.
[35] Ibid., p. 1.
[36] Cited in the *Annual Report on Tea Culture in the Province of Assam for 1882*, no. 1207, p. 5, IOR/V/24/4278, British Library, London.
[37] The following discussion has been compiled from *Annual Reports on Tea Culture in the Province of Assam*, 1883–1895 (hereafter ARTC), IOR/V/24/4278-9, British Library, London and the *Annual Reports on the Administration of the Province of Assam*, Assam State Archives (hereafter ASA), Guwahati, Assam; "outturn" here refers to amount of tea produced, or crop yield.

Table 3.1 Statistics showing tea yields per acre, percentage increase or decrease, and variation from previous years. Note that returns are not shown for all districts and yields vary greatly between regions in Assam. Compiled from *Annual Report on Tea Culture in Assam for the years 1883–1895*, Shillong: Assam Secretariat Press

Year	District	Rate of outturn per acre (in lbs.)	Total Yield (in lbs.)	Increase (+) or Decrease (−) from previous year (in lbs.)	Percentage Increase (+) or Decrease (−)
1884	Cachar	272	12,576,899	−3,38,097	−2.61
	Darrang	330	4,384,141	−1,49,012	−3.28
	Nowgong	332	3,074,115	−6,29,360	−16.99
	Lakhimpur	437	11,317,813	−1,013,008	−8.21
1885	Nowgong	314	2,805,940	−2,68,175	−8.72
	Sibsagar	338	12,854,864	−3,09,885	−2.35
1887	Lakhimpur	487	13,011,899	−3,83,892	−2.87
1888	Cachar	319	15, 477, 096	−1,079,202	−6.52
1889	Goalpara	302	92,083	−10,317	−10.08
	Nowgong	340	3,521,595	−2,41,449	−6.42
1890	Kamrup	194	1,152,086	−11,641	−1.00
	Darrang	467	8,433,809	−12,107	−0.14
1891	Kamrup	209	1,019,378	−1,32,708	−11.52
	Nowgong	310	3,375,417	−4,47,960	−11.72
1892	Cachar	310	16,506,444	−3,287,107	−16.66
	Sylhet	463	18,649,385	−1,310,052	−6.56
	Kamrup	168	7,69,384	−2,49,994	−24.52
	Nowgong	296	3,209,496	−1,65,921	−4.91
	Sibsagar	358	18,094,557	−2,370,039	−11.58
	Lakhimpur	475	15,567,207	−4,12,119	−2.58
1894	Cachar	339	18,348,061	−9,17,495	−4.76
	Kamrup	194	7,76,495	−1,58,337	−16.94
	Lakhimpur	465	17,431,270	−1,381,526	−7.34
1895	Kamrup	136	6,60,328	−1,16,167	−14.96
	Darrang	455	11,036,662	−1,537,808	−12.23

tea yields by −2.87 percent from the previous year. Damaging hail and red spider attack in 1888 decreased output in Cachar. Blights, red spider attack, damp weather and erratic rainfall were all reported to have significantly lowered tea production in 1892, and especially in the indicated districts. Outturn figures for 1894 in Cachar, Kamrup and Lakhimpur districts were noticeably less than the previous year because of finer plucking, blights, and bad weather throughout the harvest season. While the ecological underpinning and constraints of the Assam plantations need hardly be overstated, some figures are confusing and merit further elaboration. For instance, per-acre yield figures for the districts of Cachar and Lakhimpur show an upward trend between 1884–88 and 1884–87 respectively, as do those in Kamrup between 1890 and 1891. Per-acre outputs in Cachar and Kamrup between 1892 and 1894 show a similar increase. Counterintuitive at first, this rise resulted from intensive machine use, increase in labor skills, and expansion of total plantation land area in these districts even as overall percentage yields continued to fall.[38]

Pests, Planters, and the Natural World

Long criticized as unreliable mavericks lacking adequate scientific training, the persistence of pests and erratic rainfall posed an unprecedented challenge to planters in nineteenth century Assam. As indicated, metropolitan intervention in these matters, though robust after 1884, remained ad-hoc and mostly pedagogic.[39] Despite numerous handbooks, manuals, and treatises on the subject, planters in Assam were forced to share and consolidate practical experience of pest management and control with each other. Often, local measures of control and eradication were tried and implemented, even if unsuccessfully. Correspondence of the period also indicate that managers in Assam regularly exchanged ideas and sought out help from peers in Java, Kangra, Darjeeling, Ceylon, and even far-flung California. With Peal as trailblazer, these planter letters, memoirs, and articles demonstrate a keen eye for participant observation

---

[38] Admittedly, these figures and my point here might seem specious to scholars familiar with the history of the Assam tea industry. To be sure, the out-turn of Indian (especially Assam) tea never markedly declined overall. The point here is not to suggest that tea yields were *quantitatively* affected by these tea pests, but rather that it remained a *qualitative* competitor to tea production, plantation operation, and the triumphalism of agrarian expertise. For a contemporary reminder of this problem, see "Rains, pests hit tea output in State," *The Assam Tribune*, July 6, 2010.

[39] Among these, J. Wood-Mason's *Report on the Tea-Mite and the Tea-Bug of Assam* (Calcutta, 1884); M. K. Bamber's *A Textbook on the Chemistry and Agriculture of Tea: Including the Growth and Manufacture* (Calcutta, 1893); E. C. Cotes' *An Account of the Insects and Mites which Attack the Tea Plant in India* (Calcutta, 1895); David Crole's *Tea: A Text Book of Tea Planting and Manufacture* (London, 1897); Sir George Watt's *The Pests and Blights of the Tea Plant* (Calcutta, 1898); Claud Bald's *Indian Tea: Its Culture and Manufacture* (Calcutta, 1908); and E. A. Andrews' *Factors Affecting the Control of the Tea Mosquito Bug* [Helopeltis theivora Waterh.] (London, n.d., Calcutta, ITA, rpt. 1910) being some of the most important scientific investigations on the subject; more recent contributions include L. K. Hazarika, M. Bhuyan, B. N. Hazarika, "Insect Pests of Tea and their Management," *Annual Review of Entomology* 54 (2009): 267–284.

and analyses that contributed to and complemented formal know-how on the subject. The latter did not emerge in isolation as expert entomological science.[40]

As with the other factors of production, pest control measures were often prohibitively expensive or unavailable within tea districts. For instance, in response to the tea blight ravage in Assam, one Darjeeling planter suggested salt at the rate of two maunds[41] per acre to be applied four times during the plucking season. Dusting tea plants with lime was also recommended.[42] Such measures, though expedient, were not always practical. In the case of both salt and lime, planters regretted that expense restricted experimental trials, salt costing nearly a rupee per kilogram. In addition, they were rarely effective as long-term solutions, pests usually returning after a period of temporary absence.[43] The politics of profit dictated that control mechanisms that did not interfere too heavily with the pocket or plantation plan were likely to be welcome and therefore tried. For instance, labor conditions and wages had long been the bone of contention between planters, district officials and the colonial state. Apart from justifiable notoriety, it had not given the Assam plantations much else in a highly competitive labor market. Planters were therefore less favorably disposed to pest control methods (such as heavy pruning and brush fire) that demanded changes to the estate rhythm and an increase in labor-hands, working hours, and pay. Introducing lethal chemicals that destroyed pest and plant alike was a double-edged sword, and its application against the red spider was much discouraged by Peal, Wood-Mason and others.[44] Paradoxically, inter-insect rivalry often contributed to pest control and acted as natural checks to single-species dominance.

---

[40] Even Sir George Watt, MB, FLS, CIE, Member, Royal Horticultural Society of England and later Reporter on Economic Products to GOI (1887–1903) remarks that among his many sources of information and assistance were the large circle of planters "whom it was my good fortune to meet during my tours." He also mentions that "interest may be said to have been first prominently aroused in the subject of pests and blights by the late Mr. S. E. Peal's paper on 'Mosquito' or, as he loved to call it, the 'Tea Bug'. Prior to the appearance of Mr. Peal's paper it had been vaguely designated 'Blight', and was viewed as a mysterious visitation. Mr. Peal showed that it was caused by an insect," quoted in Watt, *The Pests and Blights of the Tea Plant: Being a Report of Investigations Conducted in Assam and to some Extent Also in Kangra* (Calcutta: Superintendent of Government Printing, 1898), p. 180.

[41] One "maund" was approximately 37.32 kilograms, with forty "seers" to a maund.

[42] See section on "Tea Blights and Pests" in *The Tea Cyclopaedia: Chapters on Tea, Tea Science, Blights, Soils and Manures, Cultivation, Buildings, Manufacture Etc., With Tea Statistics*, (London, 1882), pp. 34j–66.

[43] Ibid., p. 40.

[44] Though outside the scope of this paper, it is noteworthy that pesticide use in Assam tea and its contemporary impact on local habitat and ecology is a matter of much scientific debate and public concern. See B. Bhuyan and H. P. Sharma, "Public Health Impact of Pesticide Use in the Tea Gardens of Lakhimpur District, Assam," *Ecology, Environment*

Commenting on the red spider, Peal remarks: "if anything eats the spider, it will be another insect, not a bird."[45] Sometimes, the counsel was decidedly bizarre or outlandish. From California came the suggestion that shrimp shells had been exported to Chinese tea growers as manure and remedy against pests. Though unconfirmed as to its success rate, this was urged as a possible option.[46] In the face of advice, helpful or otherwise, planters regularly drew attention to pests that had gone unnoticed or were restricted to specific habitats and estates. Writing from Cachar, one planter sought peer response and remedy for a particular blight common in his garden, a large species of the *Psychida* family that Peal had reportedly forgotten or was ignorant of.[47] A little insect "of the ladybird tribe" that allegedly struck at the pekoe tip and caused it to droop was also discussed as a noteworthy omission from available handbooks and planter accounts of tea pests.[48] The ubiquity of the pest problem was not lost even in memoirs of the Assam tea plantations. Lady (Mrs.) Beatrix Scott, wife of a civil servant posted in Assam, records how Daku, a young boy from the labor lines often earned extra pennies picking off red spiders and blights from the tea plants.[49]

Despite the localized characteristics of tea pests, planters in Assam during this period looked far and wide for solutions to their everyday problems. In the process, they forged knowledge networks with peers across the imperial divide and became aware of similar concerns in competing agrarian landscapes. In one such instance, planters in Assam and Darjeeling discovered Mackenzie's first edition book on the effects of mildew, rust, and smut on North American wheat. The findings were chillingly comparable: "blight originates from moist or foggy weather and from hoarfrost, the effects of which, when expelled by a hot sun, are first discernible on the straw."[50] The depredations of the red spider on English wheat were very similar to Assam tea and Mackenzie's suggestions of control were seriously discussed. Letters from tea growers in Ceylon remarked that the effects of the monsoons and tropical

---

and Conservation Vol. 10, No. 3 (2004): 333–338 for an example. Also see the section "What of the Tea Bugs" in the Conclusion.

[45] *The Tea Cyclopaedia: Chapters on Tea, Tea Science, Blights, Soils and Manures, Cultivation, Buildings, Manufacture Etc., With Tea Statistics*, p. 39.

[46] Ibid., p. 45.   [47] Ibid., pp. 40–42.   [48] Ibid., pp. 50–52.

[49] "Daku: A Little Boy from an Assam Tea Garden," Lady B. Scott Papers, Box II, Assam 1917/1926, Given by G. P. Stewart, Center for South Asian Studies, Cambridge University, Cambridge.

[50] *The Tea Cyclopaedia: Chapters on Tea, Tea Science, Blights, Soils and Manures, Cultivation, Buildings, Manufacture Etc., With Tea Statistics*, p. 43.

weather variations were far more pronounced on their crop than in Assam.[51]

Keeping a close ear to local pest vernaculars and methods of control was also necessary under the circumstances. Under attack from a "peculiar kind of small insect," the manager of the sprawling forty-acre Ghyabaree tea estate sprayed his tea saplings with *titapani*, Assamese for a bitter and pungent concoction drawn from the neem tree (*Azadirachta indica*). He reported that the measure, though unsuccessful at the time, had wide local acceptance as an insecticide and was thought to be an effective remedy against tea pests.[52] Pest identification was a complex process, and local names and signifiers found their way in the plantation vocabulary of nineteenth century Assam. Commenting on the tea grub that left damaged stems and limbs with a pale brown appearance, one planter records that the Assamese called it *"Batea Banda Puk,"* or the insect that made its own house or cocoon.[53] Planters were periodically compelled to consult with lower-level functionaries, especially Bengali and Assamese clerks for suggestions and advice. In the deeply entrenched and clearly defined hierarchies of power in the plantations, these exchanges often upended the relationship between patron and client, master and servant. Harold Maxwell Lefroy, appointed Imperial Entomologist of India in 1905 was not mistaken when he reportedly claimed that "much may be learnt from enquiries pursued by the Mamlatdar or Tahsildar (district revenue collectors) and especially in regard to the attitude of the cultivator towards his pests."[54] Indigenous formulas of control were often strikingly innovative and managed to check insect growth. P. R. H. Longley reminisced how his "native clerk" engineered a clever trick to kill *ghundi pokas* (green beetles) in the estate rice-fields. It seems his method of deploying dead frogs on stakes, attractive as diet but fatal when consumed, worked beautifully in curbing the menace.[55] The case of the ghundi beetle, though a paddy bug, is interesting and relevant to our story. Despite being a local staple, the emergence of rice cultivation

---

[51] *The Indian Planters' Gazette and* Sporting *News*, 25 August 1885, p. 182, Asian and African Studies, Microfilm Series MFM.MC1159, British Library, London
[52] Letter to the Editor, *The Indian Planters' Gazette and Sporting News*, 21 September 1886, Asian and African Studies, British Library, London.
[53] Lady B. Scott Papers, Box II, p. 55.
[54] Quoted in J. F. M. Clark, *Bugs and the Victorians*, pp. 187–215.
[55] Longley writes: "I can only advance the hypothesis that the carnivorous diet, though tasty, is poison to the ghundi beetle," in P. R. H. Longley, *Tea Planter Sahib: The Life and Adventures of a Tea Planter in North East India* (Auckland: Tonson Publishing House, 1969), p. 108. The depredations of the red slug and the looper caterpillar are also mentioned.

in and around the tea plantations had to do with significant managerial manipulation. Dotting estate peripheries and often found alongside labor housing areas, paddy cultivation was encouraged by planters as a cheap food source and was viewed as an economic sop to enlist new and contract-expired labor.[56] Its effects on the plantation world were however not unmixed. In the next chapter, we see how this policy led to the unmanageable rise of malarial anopheles mosquitoes that in turn affected worker health in the estates. The rise of rice pests only compounded planter problems in dealing with this scourge.[57]

Interestingly, the history of tea pest management in the Assam gardens also unearths subtle but little examined transcripts of labor resistance. While more visible forms of labor protests such as physical violence and desertions have been well documented,[58] opposition often came in unexpected ways. In one such instance, planters had considerable difficulty using bone dust as tea fertilizer due to the caste regulations of workers. Animal ash being "polluting" to many, laborers struck work demanding alternative measures. This fertilizer initiative succeeded only after planters hired "coolies of low caste" whose social position permitted its use.[59]

Not everything about tea, however, could be perfected by innovations – agrarian or otherwise. Much was unknown about Assam's topography, hydrological patterns, and tea-ecology, even as lands continued to be parceled out to prospective speculators and tea companies. Consider the case of Messrs. Duncan (Brothers) and Co. around the turn of the century. Having invested in enormous swathes of wastelands for tea, they eventually discovered that the area was incompatible with planting. They petitioned the district administration to relinquish around 798 acres in 1901 and were finally granted the release in April 1902.[60] The company cited unsuitable soil conditions and unexpected flooding as two primary reasons for abandoning the property.[61] In their submission, Duncan

---

[56] See Chapter 4 for an elaboration of this issue; Peal also comments on the green beetle in his chapter on the tea mosquito bug and writes that: "I have searched in vain for cures, and the natives say that when 'Gandhi' (the rice bug) attacks the paddy, nothing can save the crop," in S. E. Peal, "The Tea Bug of Assam," p. 130.
[57] For the paddy bug, see department of Agriculture, Eastern Bengal and Assam Bulletin, No. 17, IOR/V/25/500/229, Asian and African Studies, British Library, London.
[58] See the Introduction for a discussion of this literature.
[59] *The Indian Tea Gazette*, reprinted in *The Tea Cyclopaedia: Chapters on Tea, Tea Science, Blights, Soils and Manures, Cultivation, Buildings, Manufacture Etc., With Tea Statistics*, p. 44.
[60] Letter No. Rev/831/4375 dated April 1, 1902, Revenue Files, Jorhat District Record Room, Jorhat, Assam.
[61] Petition No. 1334, dated August 26, 1901, Court of the Collector and Deputy Commissioner of Sibsagar, Revenue Files, Jorhat District Record Room Archives, Jorhat, Assam.

Brothers reported that initial costs had not accounted for extensive drainage works and soil treatment needed for any tea planting to take off. Even after six decades of its start, tea cultivation remained a flawed science. Harold H. Mann, scientific officer to the ITA reminded planters as late as 1907 that producing good tea depended on a great variety of minute factors and circumstances, some in his control but mostly outside his power and beyond even his knowledge. He argued: "ours is a unique industry, one in which we are [still] treading untrodden ground ... our knowledge [of tea planting] is as yet imperfect beyond measure ...."[62] To use Carolyn Merchant's formulation, tea-making was constantly in tension between *natura naturans* – an active, creative and potentially uncontrollable nature and *natura naturata* – the created world purportedly describable by scientific precision and "experimentation."[63] Whereas Mann's struggle to know and perfect tea manufacture continues to this day,[64] the ecological consequences of this plantation economy highlight the limits to botanical triumph and "scientific" order that tea set out to impart in eastern India.

Natural calamities added yet another challenge to the functioning of these plantations. Part of an active seismic zone, earthquakes have been common in Assam since recorded history began. The tremor of June 12, 1897 was particularly devastating and impacted plantation life and landscape significantly. W. M. Fraser recalled that the land heaved, throwing everyone off balance. It proved to be a terrifying experience for laborers and planting work effectively ceased in its aftermath.[65] The official report on the earthquake detailed huge storm surges and damaged crops, livestock, roads, and property.[66] Almost five decades later, the earthquake of August 15, 1950 caused widespread mayhem in the tea districts of Doom Dooma, Panitola, Dibrugarh, and North Lakhimpur. It led to landslides and an unprecedented damming up in higher reaches of the

---

[62] Harold H. Mann, *The Factors Which Determine the Quality of Tea*, Indian Tea Association Bulletin No. 4/1907, 2; 29, Mss Eur. F/174/1515, British Library, London.

[63] See Carolyn Merchant, *Autonomous Nature: Problems of Prediction and Control from Ancient Times to the Scientific Revolution* (New York, NY and London: Routledge, 2016), pp. 164–165.

[64] Consider a mix of headlines pertaining to tea over the past seven years from the leading English daily of the region, *The Assam Tribune*: "Super bugs threaten to eat into vitals of the industry" (March 2011); "Tea Industry Passing Through Critical Times" (November 1, 2015); "Adaptation to Climate Change in Tea Mooted" (May 28, 2016); "Rains to Impact Tea Output in State" (July 29, 2016); "All is Not Well with Tripura Tea Industry" (August 5, 2016); "Irregular Gas Supply Hits Tea Factories" (September 1, 2016).

[65] W. M. Fraser, *The Recollections of a Tea Planter* (London: Tea and Rubber Mail, 1935), p. 68.

[66] See *Report on the Earthquake of the 12th June 1897*, No. 5409G/A4282, ASA, Guwahati, India.

Dehing and Subansiri rivers. These were eventually breached, leading to widespread flooding, damage to crops and plantations, and flotsam of felled forest trees that impeded transport and inland waterways.[67] Wildfires were equally destructive. On the morning of March 7, 1867, storms fanned an uncontrollable fire that burnt down a tea house and killed another laborer who attempted to douse it.[68] Many years later, a virulent influenza epidemic in 1918 was reported to have alone caused a reduction of crops by half-a-million pounds.[69]

Consider the exasperation of a planter that effectively sums up the vexed relationship between economy and ecology in these plantations:

> Don't tell me about the benevolent order of Nature ... here I am to be sacked because rain fell for three weeks out of every four, and kept the thermometer at 68 ... the wisdom of turning managers out because the meteorology of the province has been unfavourable to the anticipated growth of tea leaf, is perhaps one of those things which my grandmother calls 'a curious *non sequitur*.'[70]

Always at the center of the pest problem, planter correspondences and memoirs suggest that empirical observations on the ground were more valuable than the discursive "fixes" of scientific manuals, handbooks, and treatises. Though widely circulated and subsequently used by the planting community, this expert metropolitan knowledge base was created with help from, and in association with, men on the spot. As it is, sociocultural histories and attitudes are embedded in the story of pests in the Assam plantations. Planters confronted labor protests when caste "boundaries" were transgressed while using fertilizers. Sometimes, indigenous methods of pest prevention and remedies were listened to and tried.

The "economics of Eden" had many obvious, and many unseen ramifications. If Assam tea filled imperial coffers and delighted metropolitan palates with its body and flavor, the enterprise's impact on the labor and landscape employed was far from benign. As beneficial host base for a plethora of tea bugs and pests, this chapter highlights one such damaging aspect of the tea monoculture ecosystem – and its self-destructive consequences on the land of its birth. To be sure, the last had not been heard of these pests despite

---

[67] See Antrobus, *A History of the Assam Company*, pp. 238–239.
[68] *Orunodoi*, March 1867, 34, in Arupjyoti Saikia, re-edited, *Orunodoi: Collected Essays 1855–1868* [in Assamese] (Nagaon: Krantikaal Prakashan, 2002), p. 440, translation mine. Originally published by the Sibsagar Mission Press, Sibsagar, Assam.
[69] Antrobus, *A History of the Assam Company*, p. 201.
[70] *The Indian Planters' Gazette and Sporting News*, November 24, 1885, Asian and African Studies, British Library, London.

advancements in agro-scientific research and techniques of planting. A recent resurgence in tea blights and their growing immunity to methods of control (whether organic or chemical) in 2011 is a bleak reminder that history continues to repeat itself in the empire's gardens.[71]

If anything, the bug problem and its attendant eco-social costs provide a telling example of the disheveled legacy of the Assam tea enterprise – and its promise of agrarian improvement and development in the region.[72]

---

[71] See report on "Super Bugs Threaten to Eat into Vitals of Tea Industry," *The Assam Tribune*, March 21, 2011.

[72] See James C. Scott's critique of State-led "development" schemes and the manipulation of nature, knowledge, and society among others in *Seeing Like a State: How Certain Schemes to Improve the Human Condition Have Failed* (New Haven, CT and London: Yale University Press, 1998), especially Chapters 1, 8 and 9.

4 Death in the Fields

> Most planters are thoroughly alive to the fact that what is in the end best for the health of the coolie is leave rather than medicine.[1]

On December 29, 1882, the steamer *Scinde* set sail with more than five hundred souls for the tea plantations of Assam in eastern British India. Despite a violent storm on the river Brahmaputra upstream, she dropped anchor at Dibrugarh the following month with her consignment of indentured labor among others. More than forty immigrants lay dead, claimed by cholera alone.[2] In the next decade, more than 92,275 tea workers perished due to cholera, kala-azar (or black-fever), malaria, anemia, dysentery, dropsy, diarrhea, respiratory disease, and "other causes" leading the list.[3] As the pattern continued well into the twentieth century and beyond, death and Assam had become almost indistinguishable.

Numbers notwithstanding, this chapter argues that ideas of health and disease in the Assam estates transcended logics of bodily disorder, pathogens, or preventive medicine. It also suggests that historical epistemologies and perceptions of mortality and morbidity – indeed of the body – in nineteenth century eastern India were seldom "scientific" or medically objective. As already indicated, law permeated every aspect of these plantations and labor life – wages, hours of work, terms of contract and health. Within this context, this chapter shows that disease etiology and epidemiology, and norms of well-being were conditioned by three concurrent expedients: medical opinion, tea profits, and regulatory

---

[1] See *Report of the Assam Labour Enquiry Committee, 1906* (Calcutta: Superintendent of Government Printing, August 6, 1906), p. 123.
[2] Reported by Dr. J. J. Clarke, Sanitary Commissioner, Assam in Annual Sanitary Report of the Province of Assam for the Year 1882 (Shillong: Assam Secretariat Printing Office, 1883), p. 34.
[3] Compiled from *Report on Labour Immigration into Assam for these years* (Shillong: Assam Secretariat Press), Assam State Archives (hereafter ASA), Guwahati, India; admittedly, these figures are conservative and belie statistical accuracy. See the third section of this chapter for details.

"standards." In other words, vector identification, prophylaxis and policy recommendations were happy to – in fact, had to – play second fiddle to commerce and costs. Often, law dictated the parameters of fitness and disease. Social histories of medicine and health in the Assam plantations therefore need to engage simultaneously with three iterations of the labor body: the pathological, the productive, and the legal.

Public health and medicine in colonial India is now a well-defined and burgeoning field. Mark Harrison's early work looked at the imperial determinants of sanitary policy, competing medical opinion between metropole and colony, nationalist response to Western therapeutics and rhetoric of reform, and the relationship between disease theory and praxis.[4] Pioneering the field of Western medicine's biosocial colonization of the "Indian" body, David Arnold argued that scientific approaches to epidemic disease control – especially smallpox, cholera, and plague – were superimposed onto ideas of corporeality and the "political concerns, economic intents, and its cultural preoccupations" of empire.[5] He thus suggests that the discursive, ideological, and institutional roles of British medical therapeutics and control both impacted and were influenced by local caste practices, gender relations, and religious norms. Arnold contends that indigenous response to this colonizing process was contingent and layered, and exceeded simplistic logics of resistance and co-option. Still others looked at the "risk perceptions" and scenarios of imperial infectious disease policy[6]; the emergence of "tropical medicine" as an ideological and empirical system of etiology and epidemiology[7]; and the systematics of colonial public health implementation through institutions, individuals, mechanisms, ideas, and "native traditions."[8] In a recent work, Nandini Bhattacharya shows that the tea plantations of northern Bengal and the hill-stations of Darjeeling and Duars provided unique eco-pathogenic sites for malarial research and idioms of (especially

---

[4] See Mark Harrison, *Public Health in British India: Anglo-Indian Preventive Medicine, 1859–1914* (Cambridge: Cambridge University Press, 1994); also, Harrison, "A Question of Locality: The Identity of Cholera in British India, 1860–1890," in David Arnold, ed., *Warm Climates and Western Medicine: The Emergence of Tropical Medicine, 1500–1900* (Amsterdam: Rodopi B.V., 1996).

[5] See David Arnold, *Colonizing the Body: State Medicine and Epidemic Disease in Nineteenth Century India* (Berkeley, CA and London: University of California Press, 1993).

[6] See Sandhya L. Polu, *Infectious Disease in India, 1892–1940: Policy-Making and the Perception of Risk* (London: Palgrave Macmillan, 2012).

[7] See David Arnold, ed., *Warm Climates and Western Medicine: The Emergence of Tropical Medicine, 1500–1900*.

[8] See Biswamoy Pati and Mark Harrison, eds., *The Social History of Health and Medicine in Colonial India* (Abingdon, Oxon: Routledge, 2009).

Western) corporal well-being, respectively.[9] She suggests, however, that imperatives of capitalist profits and politics forestalled widespread implementation of disease prevention and health care policies in these estates. In other words, Bhattacharya argues that locality – as discourse, concept, and "field" – conditioned and constrained tropical medical research and practices of public health in colonial India.

To be sure, this is not an exhaustive reading of the field, nor does it do justice to the range and depth of its analytics. However, even a summary glance shows that social histories of health and medicine in South Asia have closely followed the career of the colonial state, in areas and aspects where its hold was firmly, and unquestionably, in place. But this begins to change when we travel to the "peripheries" of empire, where the zone between formal control and laissez-faire is loosely defined. Here, competitors to colonial power and authority, and exigencies of commerce, begin to modulate notions and priorities of medical opinion, health, and morbidity. Here, "who speaks for the body of the people?"[10] can no longer be answered simply in terms of indigenous response or hegemonic metropolitan discourse. Here, the subjects of medical intervention and disease prevention are not just regimented (as in the Army and Jails) but also legislated. The "enclavist argument"[11] of medical responsibility and successful disease control in cantonments and prisons is undercut by proprietorial battles of "coolie ownership" between planters and the colonial state. Here, quantitative analyses of risk scenarios and death-rates among "new recruits,"[12] the institutional history of therapeutics,[13] or an overbearing indenture regime[14] do not fully account for *how* health and disease were implicated with law, medical discourse, and profiteering in these estates. Even Bhattacharya's "logic of locality" remains inadequate because of structural dissimilarities between the plantation economies of Assam and north Bengal.

In what follows, "political economy of health" is first used as a theoretical window to understand *why* well-being and disease exceeded – indeed had to exceed – somatic and scientific logics of disorder

---

[9] See Nandini Bhattacharya, *Contagion and Enclaves: Tropical Medicine in Colonial India* (Liverpool: Liverpool University Press, 2012).
[10] See Arnold, *Colonizing the Body*, p. 10.   [11] Ibid., p. 96.
[12] See Ralph Shlomowitz and Lance Brennan, "Mortality and Migrant Labour in Assam, 1865–1921," *The Indian Economic and Social History Review* Vol. 27, No. 1 (1990): 85–110.
[13] Achintya Kumar Dutta, "Medical Research and Control of Disease: Kala-Azar in British India," in Biswamoy Pati and Mark Harrison, ed., *The Social History of Health and Medicine in Colonial India*.
[14] Rana P. Behal and Prabhu P. Mohapatra, "Tea and Money Versus Human Life: The Rise and Fall of the Indenture System in the Assam Tea Plantations 1840–1908."

in this story. In the process, we see that the laboring body in these estates had overlapping economic, pathological, and legal subtexts that were called upon as necessary. The next two sections expand on this idea by focusing on disease and law, respectively. Etiological debates on cholera, kala-azar, and malaria – the three main killers in the Assam gardens – are discussed through the eyes of sanitarians, medical investigators and estate doctors. In doing so, I show how scientific and policy recommendations for prevention and cure were conditioned by – and very often ran counter to – the politics of profit in the tea industry. But sickness and morbidity in the Assam plantations also had legal ramifications. For law both aided commercial interests and concurrently set regulatory parameters of labor health and sickness. This chapter ends by analyzing this paradox as a peculiar feature of the Assam tea enterprise.

Political Economy of Health

Health is too important to be entrusted only to doctors[15]

The social history of health in the Assam plantations cannot be understood within the hermeneutics of somatic disorder. Nor was it exclusively a public health issue – for neither "public" nor "health" was clearly defined. Here, normality and well-being depended on maintaining an operational equilibrium between medical prophylaxis, profits, and legal regulation. In other words, the peculiar structures of the tea industry, and its dominant relations of production, have to be taken into account in understanding histories of health and impairment in Assam.

As a theoretical paradigm, the political economy of health has been used in medical socio-anthropology since at least the 1970s.[16] At its most

---

[15] Hans Peter Dreitzel, ed., *The Social Organization of Health*, Recent Sociology No. 3 (New York, NY and London: Macmillan, 1971).
[16] For an early exposition of the theory (and definition) of political economy of health, see Sander Kelman, "Introduction to the Theme: The Political Economy of Health" and Kelman, "The Social Nature of the Definition Problem in Health," both in *International Journal of Health Services*, Vol. 5, No. 4 (1975): 535–538; 625–642 respectively; also see Vicente Navarro, *Medicine Under Capitalism* (New York, NY: Croom Helm Ltd., 1976), Hans A. Baer, "On the Political Economy of Health," *Medical Anthropology Newsletter*, Vol. 14, No. 1 (Nov. 1982): 1–2, 13–17, Howard Waitzkin, "The Social Origins of Illness: A Neglected History," *International Journal of Health Services*, Vol. 11, No. 1 (1981): 77–105, Merrill Singer, "Developing a Critical Perspective in Medical Anthropology," *Medical Anthropology Quarterly*, Vol. 17, No. 5 (1986): 128–129; for more recent studies, see Howard Waitzkin, *Medicine and Public Health at the End of Empire* (Boulder, CO: Paradigm Publishers, 2011), Vicente Navarro, ed., *Neoliberalism, Globalization and Inequalities: Consequences for Health and Quality of Life* (Amityville, NY: Baywood Publishers, 2007), and Clare Bambra, *Work, Worklessness, and the Political Economy of Health* (New York, NY: Oxford University Press, 2011).

general level, this approach studied disease distribution, illness, and health-care as effected by the economic, especially capitalist, organization of society. The etiology of sickness and mortality was therefore examined as a multifactorial, multi-causal socio-economic phenomena rather than a bio-medical aberration. Of course, there were methodological variants to this way of thinking; the "orthodox Marxists" looked at the "predictable logics" of capitalist exploitation of wage-labor and its attendant impact of health and sickness[17]; the "cultural critics" saw medical practices as replicating patterns of social inequality and exclusion; and "dependency theorists" linked imperialism, colonialism and capitalist expansion to ("Third World") under-development, poverty, and disease.[18] To be sure, Foucault's history of biosocial regulation and state control can also be included here, though his study is not explicitly about disease or health per se.[19] It is not my intention to enter into the full theoretical range of these positions, nor do I uncritically subscribe to any *one* version as being more valid than the other. Instead, I argue that this theoretical perspective allows us a deeper understanding of three critical components of our story namely the body, disease, and health.[20] In the process, it clarifies the logic of law in this plantation form and its connection to these three factors in the region.

One of the most helpful aspects of this method was its nuanced deconstruction of the organismal integrity of the human body and models of health. It rejected the classical approach of health scientists that looked at the body as an automaton, a bio-machine of sorts. For them, sickness was a biologically induced malfunction of this well-tuned mechanism. The "social epidemiological school," though cognizant of the societal basis of disease, was also critiqued for its implicit acceptance of illness

---

[17] For the "Marxists," of course, Engels' 1845 treatise, *The Condition of the Working Class in England* remains the classic statement on the correlation between capitalist expansion and poor health; see Friedrich Engels, *The Condition of the Working Class in England*, trans. W. O. Henderson and W. H. Chaloner (Stanford, CA: Stanford University Press, rpt. 1958).

[18] These distinctions are used and elaborated in Lynn M. Morgan, "Dependency Theory in the Political Economy of Health: An Anthropological Critique," *Medical Anthropology Quarterly*, New Series, Vol. 1, No. 2 (June 1987): 131–154.

[19] See Michel Foucault, *The History of Sexuality*, Vol. 1, 2 and 3 (New York, NY: Vintage), and *"Society Must be Defended": Lectures at the Collège de France, 1975–1976*, trans. David Macey (New York, NY: Picador, rpt. 2003), especially lectures One and Eleven.

[20] My use of the "orthodox Marxist" approach to the political economy of health comes with several riders. For one, I argue that there was no "predictable logic" to the forces and relation of production in the Assam estates, nor did it follow specific rules. As shown, these were contingent on the structure of recruitment, notions of authority and power, and the stipulations of law. I also don't consider medical knowledge in these plantations as part of an elaborate "ideological framework." Of course, capitalism as an economic system was never the same everywhere.

as an index of "lifestyle."[21] Early proponents of the political economy approach, especially Sander Kelman and Talcott Parsons, argued that neither "school" tackled the substantive meaning of health – as condition and concept – that arose from a complex dialectic between somatic and social causes. For them, the material relations in society was an important starting point as it engendered these meanings, and an individual's relationship to it, in the first place. Thus:

> The definition of health cannot be normatively chosen on *a priori* grounds, but rather must be derived (in practice *is* derived) from the social and institutional dynamics of the society in question. In particular where expansive commodity production is combined with the control over production by a small, but powerful, social class, there is a tendency for health to become institutionally defined in functional, rather than experiential, terms.[22]

These exegetical distinctions of health need further elaboration. Under capitalism, functional health refers to an externally imposed parameter of well-being, "a state of optimum capacity of an individual for the effective performance of the roles and tasks for which s/he has been socialized."[23] Thus, instrumentally, it refers to providing for, and maintaining a state of bodily condition that does not interfere with, or diminish, the accumulation of capital. On the other hand, experiential health is more phenomenologically defined as the intrinsic, self-perceptual understanding of wholesomeness, freedom from illness, "the capacity for human development" and the "transcendence of alienating social circumstances."[24] In other words, this is a state of wellness that is *experienced* rather than *standardized*. To be sure, these distinctions are theoretical rather than medical and are not mutually exclusive. Kelman clarifies: "in sum, experiential and functional health represent two qualitatively different notions or norms of organismic integrity which are either promoted or stunted in different forms of society."[25] From the perspective of philosophical anthropology, these norms also relate to different corporeal states –

---

[21] See Kelman, "The Social Nature of the Definition Problem in Health," especially pp. 628–634.
[22] Kelman, "Introduction to the Theme: The Political Economy of Health," p. 537; also, Talcott Parsons, "Definitions of Health and Illness in the Light of American Values and Social Structure," in E. Gartly Jaco, ed., *Patients, Physicians, and Illness: A Sourcebook in Behavioral Science and Health*, Third Edition (New York, NY: The Free Press, 1979), pp. 120–144.
[23] Parsons, "Definitions of Health and Illness," p. 132; also, Kelman, "The Social Nature of the Definition Problem in Health," pp. 629–634; Hans A. Baer, "On the Political Economy of Health," p. 14.
[24] Kelman, "The Social Nature of the Definition Problem in Health," p. 629.
[25] Ibid., p. 630.

functional well-being is about "having a body" while experiential health draws from "being a body."[26]

Theoretical debates aside, these discussions of health and the body have important historical relevance to our story. The health of the Assam plantation laborer, as means of production, was a functional concern; her/his role performance defined by the operation of these estates and the logic of capital.[27] Malarial investigations, inoculation drives, etiological studies, prophylaxis, and prevention continually negotiated well-being with an eye to costs and profits. But beyond this, the Assam plantations had important structural peculiarities. As an indentured worker, she/he (the Act-Labor in this case) was also the subject of law.[28] The regulation of sanitary welfare, indeed health, was an inextricable and important part of labor legislations. In this respect at least, functional health in Assam was not just about maintaining productive efficiency; it was also dictated by juridical stipulations, lapses notwithstanding. With Act-Labor particularly, sanitarians, medical men, and planters were therefore simultaneously dealing with legal and economic "bodies" in terms of scientific experimentation, therapeutics, work schedule, physical capacity, and standards of wellness. It was in this context that a statistical deviation of 7 percent made the difference between "healthy" and "diseased"; where, theoretically, law could both cause and arrest mortality in the gardens; where the labor protector could be expected to fulfill both legal and medical roles; and where the paradox of fitness and productivity operated. Reading into these correlations between commerce and regulation colored attitudes to health and mortality in this history – in them we locate our first two bodies of evidence.

There was a third dimension of health and disease in the Assam plantations. This related to the natural, organic body of the laborer stripped of its legal and economic functions. She/he not only generated capital in the tea estates, and signed agreements, but also *lived* sickness and well-being. For purposes of disease and health, this body was referred to in humoral, ecological terms. Here, morbidity was an attribute of the naturally insanitary, indolent, non-immune "coolie" or the "weakly

---

[26] See Edmund Husserl, *The Idea of Phenomenology*, trans. William Alston and George Nakhnikian (The Hague: Nijhoff, 1964), Jean-Paul Sartre, *Basic Writings*, ed. Stephen Priest (London: Routledge, 2001), especially chapter 3.

[27] To be sure, "experiential" health in terms of the Assam laborer is almost impossible to grasp and narrate – even conceptually – for she/he only *appears* in archives and records as transcribed and represented for.

[28] By "Act-Labor" were meant those who signed contracts under the prevailing labor laws; see the fourth section of this chapter for an elaboration of this point.

Behari."[29] The "bad-batch" theory of disease endemicity and epidemics, invoked by doctors, sanitary commissioners, and planters throughout this period used this third corporal register. Under this rubric, the body fell ill due to germs, habit, lifestyle, class and racial characteristics, location, or processes of imperfect modernization depending on points-of-view. Shorn of the logic of capital, and law, our third body of evidence was read in its visceral, natural, biological state. Not infrequently, these three understandings of the body and health clashed with each other, as we shall see.

Routes of Infection and Roots of Disorder: Cholera, Kala-Azar, and Malaria

On November 1899, a severe cholera outbreak killed hundreds of laborers working in the Lumding section of the Assam Bengal Railways (hereafter ABR). Shortly thereafter, a war of words broke out between the Agent of the Company and the Chief Commissioner of Assam, Mr. H. J. S. Cotton. Charges and counter-charges centered around two issues: the party responsible for the infection and its dispersion, and financial liability for welfare arrangements undertaken.

In the immediate aftermath of the Chief Commissioner's opprobrious message to ABR for failing to effectively combat cholera, both the Consulting Engineer to GOI for Railways, Assam and the Agent and Chief Engineer, ABR were quick to focus on the question of infection. Here, they were keen to dispel notions that cholera was contracted on railway territorial boundaries. Arguing that laborers for the ABR were recruited from "districts free of cholera,"[30] the Consulting Engineer squarely blamed previously infected areas in Assam through which they passed for the disease. Echoing his colleague, the Agent and Chief Engineer argues that cholera could not have originated and spread in ABR jurisdiction, or through steamers used by its laborers, being duly under the charge of "fully qualified [European] Medical Officer, Assistant Surgeons, and Hospital Assistants" in every division. He hastens to add:

---

[29] The relationship between ethnology and labor suitability in colonial mines, factories, and plantations in India have been a recurring, and well-studied, historical debate; see Kaushik Ghosh, "A Market for Aboriginality: Primitivism and Race Classification in the Indentured Labour Market of Colonial India," in Gautam Bhadra, Gyan Prakash, and Susie Tharu, Eds. *Subaltern Studies X: Writings on South Asian History and Society* (New Delhi: Oxford University Press, 1999), pp. 8–48; also see Jayeeta Sharma, "'Lazy Natives,' Coolie Labour, and the Assam Tea Industry," *Modern Asian Studies*, Vol. 43, No. 6 (2009): 1287–1324.

[30] See Assam Secretariat Proceedings, General Department, Home A, September 1901, No. 58–66, ASA.

Death in the Fields 105

"there was no epidemic cholera at Lumding, but that it was all brought from Gauhati."[31]

Epidemiologically, the question of routes and roots of cholera infection amongst ABR workers was connected to accepting financial responsibility for segregation camps, rest houses, and hospitals established for their cure. In fact, the Chief Commissioner of Assam passed orders to this effect on November 6, 1899. Unsurprisingly therefore, the debate here on "proving" the origins of cholera and the general unhealthiness of ABR workers thereof centered not on "bad batches" or on contagionist theories of bodily disorder. Waiting for unclaimed bills to be paid off, ABR's Executive Engineer wrote angrily to Assam Chief Commissioner wondering why they should be made to pay for sanitary and medical arrangements used not by "railway coolies" but presumably by their counterparts in the tea estates.[32]

As it is, the etiological career of cholera in India was a long and vexed one. Designated variously as an "Asiatic disease," a disease of filthy Hindu pilgrims, and one rooted in poverty, colonial medical readings of cholera vacillated between the contagionist, atmospheric, and racial theories of bodily disorder.[33] Dr. James L. Bryden (1833–80), the first statistical officer of the then newly established Sanitary Department of GOI, expressed himself strongly in favor of the airborne theory of cholera dispersion, occurrence and etiology. Finding acceptance among powerful figures in the Indian Medical Service (hereafter IMS), especially J. M. Cuningham (1829–1905), India's longest-serving sanitary commissioner, the anti-contagionist lobby for long stemmed the tide against those who called more direct intervention in matters of public health. Chief among them was Annesley Charles C. DeRenzy (1828–1914), once sanitary commissioner of the Punjab and a consistent and vociferous partisan of the water-borne theory of cholera. While the debate between DeRenzy and the GOI is well known,[34] it is important for our story that he was removed from a civil to a military position in "remote Assam" for daring to challenge the

---

[31] Ibid., p. 3.
[32] Ibid., p. 4; the Executive Engineer goes on to suggest: "this cholera would not have been contracted on this railway, and one might deign to presume that the people bringing them into the district in such a condition should bear any expenditure incurred on their account," vide letter no. 251E dated January 4, 1900.
[33] See Arnold, "Cholera: Disease as Disorder," in David Arnold, *Colonizing the Body: State Medicine and Epidemic Disease in Nineteenth Century India*, pp. 159–199; and Mark Harrison, "A Question of Locality: The Identity of Cholera in British India, 1860–1890," in David Arnold, Ed. *Warm Climates and Western Medicine: The Emergence of Tropical Medicine, 1500–1900*. See also Mark Harrison, *Public Health in British India: Anglo-Indian Preventive Medicine, 1859–1914*.
[34] See Harrison, *Public Health in British India*, especially pp. 135–146.

"official" line on the disease.³⁵ He was to subsequently take on the role of Assam's sanitary commissioner in 1877, a post he held till 1880.

Historians have remarked that the difference in medical opinion on cholera in late-nineteenth-century India was a battle of egos, a territorial fight to secure the originary, yet distinctive character of the disease versus its European iterations. While the minutiae of this discussion need not detain us here, it is significant to note that, given the post-Mutiny anxiety of the government in meddling in "native" affairs, the interventionist attitude to cholera prevention, prophylaxis and cure took somewhat of a backseat till the early twentieth century. David Arnold thus argues: "the tenacity with which so many medical researchers and senior medical advisers like Cuningham clung to their anti-contagionism can be seen as an indication of how far out of touch they were with advances in medical science in Europe. But they were certainly encouraged by the 'ostrich-like' mentality of the Government of India, which preferred for political and commercial reasons to pursue a noninterventionist, laissez-faire policy towards cholera."³⁶ Of course, as Arnold and Harrison both show, this mindset slowly changed post Robert Koch's "discovery" of the comma bacillus organism in water in Calcutta in 1884 though the effective "enthronement" of the bacteriological origins of cholera in India was to take place only in the first decades of the twentieth century.³⁷ To be sure, Waldemar M. Haffkine, the Jewish-Russian bacteriologist who pioneered anti-cholera inoculation in India during his experiments in 1893–96 was much mistrusted by the anti-contagionist camp and was even suspected by some newspapers of being a spy.³⁸ Despite continuing difference of opinion regarding the source of cholera infection, Haffkine's vaccination experiments took him Sialkot, Hardwar, Calcutta, Gaya, and Darbhanga jails in Bihar and to forty-five intensely choleraic tea plantations of Assam in 1894–5.³⁹ In his official report to the GOI, Haffkine writes:

I would think it most important that Government should recommend the inoculations to large bodies of men under their charge who, in the course of their duties, are subject to cholera epidemics, such as the troops, prisoners, coolies employed on railways, on public and military works, coolies passing through emigration depots, etc . . . A [circular] to large employers of labour, such as the directors of tea companies, coal mines, private railway lines, etc., drawing their attention to the

---

³⁵ Ibid., p. 103.   ³⁶ Arnold, *Colonizing the Body*, p. 195.   ³⁷ Ibid., p. 195.
³⁸ Pratik Chakrabarti, "Curing Cholera: Pathogens, Places and Poverty in South Asia," *International Journal of South Asian Studies*, Vol. 3 (December 2010): 153–68.
³⁹ See W. M. Haffkine, *Protective Inoculation Against Cholera* (Calcutta: Thacker, Spink & Co., 1913), especially Part II.

Death in the Fields 107

inoculations would help in diffusing knowledge on the subject and aid in the progress of this work.[40]

Haffkine's suggestion raises important questions regarding public health, colonial state intervention, and the discourse of efficiency. While Haffkine's ideas might have been considered suspect by the ilk of Cuningham and others, the issue here was not about medical opinion alone. Using the political alibi of non-interference, the colonial government largely resisted compulsory vaccination as a public health measure almost till the 1930s. For economically productive bodies such as prisoners, soldiers, and tea estate workers, the prophylactic benefits of inoculation surpassed political considerations. Instructions were duly issued by the Government of Bengal to accord Haffkine all support in his endeavors, and especially in the endemic "reservoirs" of the disease in deltaic northeastern India.[41] Of course, as this chapter shows, planters repeatedly resented (and denied) the natural reservoir theory of cholera's origins to argue that they were either "imported" by, acquired en route, or were inherent in the racial makeup of workers. In Assam, Haffkine faced an additional problem. If vaccination was necessary, its attendant side effects, namely incapacitating fever, reduced labor time in the estates.[42] Haffkine was thus forced to relocate the field of his demonstrations to a more stable, and controlled target group in Bihar and Bengal.

To be sure, cholera left a trail of devastation and death in eastern India throughout the period under review. By one estimate, mortality rates due to cholera in tea laborers between 1871 and 1878 averaged around 47.8 percent of the total workforce.[43] Even in the turn of the century between 1901 and 1920, cholera alone was responsible for around 13.3 percent of labor death in Assam.[44] Despite these numbers, there was no consensus among medical men, planters, and sanitary commissioners regarding cholera etiology and infection. In a well-known narrative, the atmospheric route, water-borne route and "bad batch" theory all jostled for acceptance depending on scientific ideologies and prevailing opinion. For Assam, what is interesting and peculiar was that these positions varied depending on cost considerations, planter lobbying, and labor law

---

[40] W. M. Haffkine, *Anti-Cholera Inoculation: Report to the Government of India* (Calcutta: Thacker, Spink & Co., 1895), pp. 49–50.

[41] For further elaboration on the "reservoir" theory, see Haffkine, *Anti-Cholera Inoculation* and Chakrabarti, "Curing Cholera," especially pp. 164–165.

[42] Chakrabarti, "Curing Cholera," p. 156.

[43] See note by Surgeon-General A. C. C. De Renzy on: "Cholera Among The Assam Tea Coolies," *The Lancet* (April 11, 1891): p. 823.

[44] See Ralph Shlomowitz and Lance Brennan, "Mortality and migrant labour in Assam, 1865–1921," *The Indian Economic and Social History Review* Vol. 27, No. 1 (1990): 105.

stipulations. In other words, though a political disease, as Arnold puts it,[45] cholera in Assam carried additional semantic burdens – namely legal, economic, and structural.

DeRenzy helps us understand this point a little better. Almost immediately after assuming charge as Assam's sanitary commissioner in 1877, DeRenzy set out to prove that the lack of clean and adequate water supply on board steamers, labor depots, and encampment grounds was at the root of the province's cholera problem.[46] He urged the government to act on these grounds and remedy defects. Debunking theories of local endemicity, imported infection, and "widespread atmospheric influence that hangs over the great river Brahmaputra,"[47] as likely etiologies of the cholera pathogen, DeRenzy undertook a somewhat covert journey aboard a steamer hauling laborers to demonstrate his point. He writes in his 1878 sanitary report (hereafter ASR) that there was but one tap providing clean water for a group of around four or five hundred laborers, and tubs filled with the cholera virus and fecal matter were thus used in its place.[48] The water of the Brahmaputra could not be blamed as it was regularly hauled and used by the crew and European staff without any perceptible inducement of cholera. While some improvements regarding water supply onboard vessels were introduced around 1879, heeding to all three of his recommendations was thought unnecessary and financially forbidding. Of course, Koch's "discovery" was still many years away, and DeRenzy's previous spat with the GOI and Bryden ensured that its response was unfavorable and haphazard at best. DeRenzy later reminisced: "it is, I think, greatly to be regretted that Government did not comply with my earnest request to make the experiment complete, as I believe it would have thrown invaluable light on many obscure points relating to cholera. Emigration *via* Brahmaputra is assuming immense proportions. Last year 20,000 coolies passed up the river. This great tide of human life affords unequalled opportunities for the exact study of cholera; but they are not turned to account because Government are taught to believe that the spread of cholera is governed by general influences which are altogether beyond human control."[49]

---

[45] Arnold, *Colonizing the Body*, p. 159.
[46] See *Annual Sanitary Report of the Province of Assam for the Year 1877* (Shillong: The Assam Secretariat Press, 1878), pp. 20–21.
[47] De Renzy, "Cholera Among The Assam Tea Coolies," p. 823.
[48] See *ASR 1878*, especially sections IV and VII.
[49] See note by Surgeon-General A. C. C. De Renzy on: "The Prevention of Cholera," *The Lancet*, August 9, 1884, pp. 227–8; by "experiment," DeRenzy refers to changes in water-supply provisions aboard steamers, at labor depots (especially) Dhubri, and at encampment sites en route. He quotes figures both "previous to the change" and "subsequent to the change" to prove falling mortality rates. For instance, as per his note,

In addition to the atmospheric theory of the miasmatic Brahmaputra, these general influences also included the German hygienist Max von Pettenkofer's "sub-soil" water theory of cholera that attributed local soil conditions and characteristics as likely agents of the disease.[50] But scientific correctness was the last thing on DeRenzy's mind when it came to his own business interests. In January 1894, a committee instituted to examine persistent unhealthiness among workers in the neighboring collieries of the Assam Railways and Trading Company (hereafter ARTC) harshly criticized the insanitary conditions in and around the mines and recommended that further labor recruitment be immediately stopped. DeRenzy led a deputation on behalf of the ARTC to counter these charges and blamed the presiding sanitary commissioner of the committee for his "foolish and impracticable advise."[51] In a letter dated February 10, 1894, DeRenzy was vocal in suggesting that the ARTC was instrumental in bringing numerous benefits to the province, and to the tea industry in general. He further argued that labor unhealthiness at ARTC was due to extraneous factors, "bad coolie importations," and "battles with nature" that the company had to contend with since operations began in 1885.[52] He also cautioned the Government of Assam that it would be ruinous for the province if its "blighting policy" of shutting down companies due to occasional labor unhealthiness continued. It is not widely known that DeRenzy was ARTC's director between 1885 and 1914.

As the debate on etiology continued, Dr. J. J. Clarke, Sanitary Commissioner of Assam, argued in 1883 (using Pettenkoffer's thesis) that the continuing incidence of cholera among Assam tea labor had to do with the 'unhygienic' conditions of areas they traversed en route to these estates. It was on account of the insanitary habits, and conditions, of these outlying local "publics" that the tea laborers repeatedly fell victim to cholera. Clarke felt that the duration of around twelve to fourteen days

---

cholera death per cent of labor population prior to these changes in 1878 stood at 2.70 whereas for 1879, 1880, 1881, 1882 and 1883 it remained close to 0.81, 0.42, 0.88, 0.65 and 0.65 percent respectively, ibid., p. 227.

[50] See Harrison, "A Question of Locality: The Identity of Cholera in British India, 1860–1890," pp. 143–144 for an elaboration of this point.

[51] See "Unhealthy Conditions of Ledo, Tikak, Namdang, and Margherita," File No. 544/5030-Emigration, District Record Room, Office of the Collector and Deputy Commissioner, Dibrugarh, Assam.

[52] Vide Letter No. 214G, dated February 10, 1894 addressed to R. S. Greenshields, Esq., Deputy Commissioner of Lakhimpur, ibid. For a general history of the ARTC, see W. R. Gawthrop (compiled), *The Story of the Assam Railways and Trading Company Limited, 1881–1951* (London: Harley Pub. Co. for the Assam Railways and Trading Company, 1951); also see Arupjyoti Saikia, "Imperialism, Geology and Petroleum: History of Oil in Colonial Assam," *Economic and Political Weekly*, Vol. XLVI, No. 12 (March 19, 2011): 48–55.

taken to cover the distance between the port of debarkation and the tea estates was too long and made workers susceptible to these choleraic influences. He argued creatively for faster steamboats that could cover the same in four to six days. In his opinion:

> A quick transit for coolie emigrants from Dhubri to the several district coolie depots is what is required. And it is to a fast transport service that we must look as a probable *solution* to the difficulties which periodically recur in regard to the epidemics of cholera on board river steamers so destructive to human life and so ruinous to the tea industry.[53]

Despite the eventual ascendancy, and acceptance, of the germ theory of disease in the twentieth century, the world of the Assam plantations continued to read them under multiple possibilities. Colored by the politics of profit, scientific propriety was only one of these many concerns. Well into the 1920s, the debate on water supply, latrines and conservancy had not been solved and medical men were reminded that recommendations should be made keeping in mind that "tea gardens live in a world of competition."[54]

Kala-Azar (or "Black Fever") had an even more complex and confusing career in the Assam plantations. Febrility was a vague and all-encompassing phenomenon and included kala-azar, malaria, and "jungle fever" in enumerative nomenclature. Especially till the turn of the century, beri-beri, ancylostomiasis (or hookworm disease), anemia, dropsy, malaria cachexia, and "all other fevers" would appear interchangeably when precise etiology was in doubt or unavailable. While debates on kala-azar research, causation, and spread is not the concern of this present chapter, suffice it to note here that, along with cholera, black-water fever remained a major killer in the Assam tea estates throughout the period under study.[55] With a plethora of vernacular names such as "saheb's disease" (or the disease of the rulers), "Blacktown fever," and "sarkari bemari" (British Government disease),[56] kala-azar confounded planters, government officials, and medical men alike.

---

[53] Vide letter dated May 19, 1883, *ASR 1882*, p. 35, emphasis mine.
[54] Question No. 97 to Dr. J. Moncrieff Joly, MB, Medical Officer, Pabhojan Tea Company, Doom-Dooma Tea Company, Tara Tea Company, in *Evidence Recorded by the Assam Labour Enquiry Committee, 1921–22* (Shillong: Assam Secretariat Press, 1922), p. 121, ASA.
[55] While figures for kala-azar mortality usually appeared under the aggregate category, "fevers" before 1891, a breakdown shows 9937 deaths in 1891, 10,247 deaths in 1893, 13,164 deaths for 1894 due the disease alone; see the respective *ASRs* for these years for an assessment.
[56] See Leonard Rogers, *Report of an Investigation of the Epidemic of Malarial Fever in Assam or Kala-azar* (Shillong: Assam Secretariat Printing Office, 1897), 132, quoted in Achintya

Figure 4.1 Charingia, Assam, India: kala-azar patients; a group of men, women, and children © Wellcome Library

Questions of endemicity, etiology, and location aside, kala-azar was an expensive disease, continually wiping out large numbers of tea laborers in Assam and eastern Bengal. Though first "identified" in the sanitary report of 1883 as a "malarial cachexia,"[57] research on the disease was to take on a long and arduous route to identification and preventive prophylaxis. It is pertinent to note here that Sir Leonard Rogers (1868–1962), IMS and professor of pathology at the Calcutta Medical School, actively sought endowments from the tea, jute, and mining establishment for his proposed Calcutta School of Tropical Medicine (hereafter CSTM) that opened doors in

Kumar Dutta, "Medical research and control of disease: Kala-azar in British India," in Biswamoy Pati and Mark Harrison, ed., *The Social History of Health and Medicine in Colonial India* (Abingdon, Oxon: Routledge, 2009) especially pp. 96–99; see also, Leonard Rogers, *Fevers in the Tropics, their clinical and microscopical differentiation, including the Milroy lectures on kala-azar* (London: Oxford University Press, 1908).

[57] Quoted in Rogers, "On the Epidemic Malarial Fever of Assam or Kala-Azar," *Medico-Chirurgical Transactions* Vol. 81, No. 1 (1898): 241–258.

1921.[58] In fact, it has been argued that the "original impetus" for research in tropical medicine (and an institute such as CTSM) came not from Rogers but from Dr. Alfred McCabe-Dallas, a young medical practitioner based in the Assam plantations.[59] Much before the mid-twentieth century, of course, the IMS had already started dispatching its officers to the tea estates of Assam in search of information and cures of this beguiling affliction. D. D. Cunningham, first sent to Assam, returned without being able to distinguish kala-azar from malaria. Surgeon George M. Giles was sent on special duty to Assam in 1889 to study the disease and made the confusing, and erroneous, claim that ancylostomiasis and kala-azar were both caused by the same agent.[60] In 1896, Rogers traveled to Assam and was followed by Sir Ronald Ross in 1898. Both their views still bordered on identifying black-water fever as a malarial epidemic.[61] Much water had to flow under the bridge before William Leishman and Charles Donovan separately identified a protozoan parasite in 1903 as the causative agent of black-water fever. In the meantime, official malariological work under the auspices of the GOI had already waded into kala-azar territory since the turn of the century. Captain S. R. Christophers of the IMS, who had participated in the work of the Malaria Committee of the Royal Society to India in 1902, collaborated with C. A. Bentley six years later to investigate the ravages of malaria and black-water fever in the Duars region of northern Bengal. Instituted after repeated petitions by Duars planters, their first report of 1908 also touched upon neighboring Assam and its plantations. Retaining their opinion on the etiological similarities of malaria and kala-azar, they proffered the additional hypothesis that its endemicity in eastern India (and especially on

---

[58] Rogers argued that in return for an investment of Rs. 20,000 for five years, the CSTM would be able to provide better answers for the many diseases affecting labor in the plantations, factories, and mines, "thereby saving many more rupees in inefficient and lost labour," in Helen Power, "The Calcutta School of Tropical Medicine: Institutionalizing medical Research in the Periphery," *Medical History* Vol. 40 (1996): 197–214.

[59] See G. C. Cook, "Leonard Rogers KCSI FRCP FRS (1868–1962) and the founding of the Calcutta School of Tropical Medicine," *Notes and Records of the Royal Society*, Vol. 60 (2006): 171–181.

[60] See G. M. Giles, *A Report of An Investigation Into the Causes of the Diseases Known in Assam as Kala-Azar and Beri-Beri* (Shillong: Assam Secretariat Press, 1890), especially section XI. Giles also added to the medical bewilderment by introducing "beri-beri" to the list. On Beri-Beri, and its medical history in India, see David Arnold, "British India and the 'Beri-Beri' Problem, 1798–1942," *Medical History* Vol. 54 (2010): 295–314.

[61] See Rogers, "On the Epidemic Malarial Fever of Assam or Kala-Azar."

Figure 4.2 A kalar-azar treatment center, Charingia, Assam, India: Indian patients are grouped outside a row of grass-roofed buildings, 1910 © Wellcome Library

its plantations and mines) had to do with the "industrial aggregation" of large number of non-immune labor at these sites.[62] Of course, as is well known, Christophers was a vocal proponent of quinine prophylaxis[63] though race, locality, and ecology also appeared as contingent factors in his discussion of the disease. As with conservancy and water supply, costs associated with aggressive quinization meant that its implementation was erratic in the plantations of Assam.

---

[62] Captain S. R. Christophers and C. A. Bentley, "Blackwater Fever," in *Scientific Memoirs by Officers of the Medical and Sanitary Departments of the Government of India*, New Series, No. 35 (Simla: Government Monotype Press, 1908), especially pp. 40–47.

[63] Ironically, they were to later "grudgingly" acknowledge that uncontrolled and irregular quinization often exacerbated incidence of kala-azar and the advanced stage of malaria known as haemoglobinuria; see Harrison, *Public Health in British India*, pp. 162–163.

As it is, Surgeon Giles made some interesting revelations about sanitary welfare in the Assam gardens despite his confusing hypothesis of kala-azar as ancylostomiasis. He suggests that one reason why estates fail to fall off the "unhealthy"[64] list despite improvements, especially in water supply, is because of their "imperfect" approach to sanitation. Arguing that ancylostomiasis was a bigger killer than cholera in the Assam plantations, Giles contends that a more robust approach to conservancy would naturally lead to a decrease in mortality figures. He rejects "theoretical considerations of the innate dirtiness of the coolie character" and says that by failing to provide for proper latrines, planters have "neglected to give his labourers the chance of becoming cleanly."[65] He goes to the extent of saying that some degree of "legal coercion" would have to be applied by planters if the scourge of ancylostomiasis was to be tackled among tea workers in Assam. Unwittingly, and from another perspective, Surgeon Giles was repeating the wisdom of the labor laws that laid down minute and specific instructions of sanitary welfare for plantation management to follow. In other words, Giles raises important points about the relationship between health, disease, and the politics of profit. We shall return to this in the next section. Closing his report to the government, Giles provides an alternative theory:

Tea-garden sanitation has, in fact, been commenced at the wrong end, for watersupply might, with comparative impunity, have been left to take care of itself, provided measures had been taken to put a stop to its pollution by means of adequate measures of conservancy.[66]

Lieutenant-Colonel T. C. McCombie Young of the IMS, previously Assam's Director of Public Health, makes an ever more forthright claim about sanitary investment and returns. In an address to the Royal Society in 1924, Young looked back at his work with kala-azar prevention in Assam and made some categorical observations.[67] He accepts Rogers' earlier suggestion on segregation for kala-azar prevention in Assam[68] but

---

[64] See the next section of this chapter for a discussion of the classificatory logic behind "healthy" and "unhealthy" tea gardens in Assam.

[65] See Giles, *A Report of An Investigation Into the Causes of the Diseases Known in Assam as Kala-Azar and Beri-Beri*, 155; also see "Surgeon Giles's Report on Anemia of Coolies," Assam Secretariat Proceedings, Home-A, No. 1–8, July 1890, ASA, especially pp. 2–4.

[66] Ibid., p. 156.

[67] See Lieutenant-Colonel T. C. McCombie Young, "Fourteen Years' Experience with Kala-Azar Work in Assam," *Transactions of the Royal Society of Tropical Medicine and Hygiene*, Vol. 18, No. 3 (June 19, 1924): 81–86.

[68] See Leonard Rogers, "The Epidemic Malarial Fever of Assam, or Kala-Azar, Successfully Eradicated from Tea Garden Lines," *British Medical Journal*, September 24, 1898, Vol. 2 (1969): 891–892, and J. Dodds Price and Leonard Rogers, "The Uniform Success of Segregation Measures in Eradicating Kala-Azar from Assam

argues that checks on labor movement, compulsory removal of infected communities, and destruction of property had not stopped it from devastating tea labor in the Sibsagar district of upper Assam in 1917. More importantly, Young suggests that this policy, costing around £20 per family, rendered it financially impossible to implement on a large scale.[69] It is not clear from his account how Young arrived at this precise figure. Withholding full confidence in the "insect vector" theory of kala-azar transmission, McCombie Young agrees that Rogers' tartar emetic solution and U. N. Brahmachari's urea stibamine, thought to be proven as universally effective in Assam, were still mitigating the spread of the disease to a great extent.[70] For him, however, the economy of health was not a matter of medical discourse alone. Its long-term financial implications were equally if not more important:

The total number of kala-azar cases treated since 1920 is now well over 80,000. Assuming a 90 per cent. success-rate in treatment, at an estimated cost of 13s. 4d. per head, I calculate that the value to the State in land revenue alone of the lives saved by these operations will, in twenty years' time, be £192,000, as against an expenditure of £53,333.[71]

Not coincidentally, perhaps, McCombie Young argued in 1922 that abolishing the breeding grounds of anopheles in the Assam estates would cost only a fraction of the money spent on quinine administration, "an expenditure [that] would yield a handsome return in an increased efficiency of labour forces."[72] Taking serious note of Young's suggestion, the Indian Tea Association funded an official investigation on malaria in the Assam and Duars tea gardens through the CSTM in 1926.

Malaria carried a heavy etiological baggage in colonial India. Long after Sir Ronald Ross's 1897 discovery of the anopheline vector in malarial transmission, debates continued on its character as a disease of place, ecology and modernization. Questions of detection, prevention, and cure oscillated between the laboratory, the "field," and the human body.[73]

Tea Gardens: Its Bearing on the Probable Mode of Infection," *British Medical Journal*, Vol. 1 (February 7, 1914): 285–289.

[69] Ibid., p. 85.

[70] Rogers first introduced tartar emetic as a treatment in 1915, while Brahmachari of the Bengal Medical Service came up with urea stibamine in 1921; see Achintya Kumar Dutta, "Medical research and control of disease: Kala-azar in British India," pp. 100–102 for a discussion of these discoveries and its impact.

[71] T. C. McCombie Young, "Fourteen Years' Experience with Kala-Azar Work in Assam," p. 87.

[72] Quoted in C. Strickland and K. L. Chowdhury, *Abridged Report on Malaria in the Assam Tea Gardens: With Pictures, Tables and Charts* (Calcutta: Indian Tea Association, 1929), p. 2.

[73] The recent historical literature on malaria in India is vast; see Michael Worboys, "Germs, Malaria, and the Invention of Mansonian Tropical Medicine: From 'Disease in the

Chief among these "fields" were the plantations of eastern India, and Nandini Bhattacharya argues that malarial research received constant patronage and entrepreneurial support from the tea agency houses throughout the late nineteenth and early twentieth century. Of course, as Bhattacharya shows, this was a paradoxical relationship as recommendations for effective malarial prevention (including site selection, non-immune labor regulation, and quinization) clashed with the logic of production and profit in these estates. The situation in Assam bore striking similarity, at least in this respect. Let us look into this a little closely.

Four years before Dr. C. Strickland, medical entomologist at the CSTM, submitted his formal report of malaria investigations referred to above, preliminary findings were read before the Jorhat branch of the British Medical Association on March 2, 1925. Strickland highlighted the need to fully understand the situation *before* adopting malaria control, as not all species of anopheline mosquitoes bred in similar conditions. He warns that ignorance, conjecture and mismanagement of malarial control measure may in fact lead to a worsening, if not proliferation of carrier species of anophelines carrying the malarial parasite:

take for instance, open-earth drainage: one drains a swamp and eradicates the species *umbrosus*; one drains another and introduces *maculatus* or *funestus*.[74]

He isolated the *funestus*, *aconitus*, and *culcifacies* as the three most dangerous anopheline species, found near streams, irrigating channels of the rice-fields, and pools of pure water during or after the rains. In addition to quinine prophylaxis and prevention, Strickland recommended species-specific measures in controlling the disease in Assam. Here, he ran into another set of structural paradoxes specific to these plantations. Suggestions against open-earth drainage proved inconsistent with agrarian technology needed for tea bush irrigation. Similarly, oiling, proposed

---

Tropics' to 'Tropical Diseases'" in David Arnold, ed., *Warm Climates and Western Medicine: The Emergence of Tropical Medicine, 1500–1900*; Nandini Bhattacharya, "The Logic of Location: Malaria Research in Colonial India, Darjeeling, and Duars, 1900–30," *Medical History* Vol. 55 (2011): 183–202; Arabinda Samanta, *Malarial Fever in Colonial Bengal: Social History of an Epidemic, 1820–1939* (Kolkata: Firma KLM, 2002); Ira Klein, "Development and Death: Reinterpreting Malaria, Economics, and Ecology in British India," *The Indian Economic and Social History Review* Vol. 38 (2001): 147–179; Sandhya L. Polu, *Infectious Disease in India, 1892–1940: Policy-Making and the Perception of Risk* (London: Palgrave Macmillan, 2012), especially chapter 3, and Elizabeth Whitcombe, "The Environmental Costs of Irrigation in British India: Waterlogging, Salinity, Malaria," in David Arnold and Ramachandra Guha, eds., *Nature, Culture, Imperialism: Essays on the Environmental History of South Asia* (New Delhi: Oxford University Press, 1997).

[74] See C. Strickland, "The Mosquito Factor in the Malaria of Assam Tea Gardens," reprinted from *The Indian Medical Gazette*, Vol. LX, No. 11 (November 1925), p. 2.

as a check against pool breeders, proved futile because of high costs, fast flowing rivers and rice cultivation in and around these estates. Furthermore, Strickland's rather unconventional argument against jungle clearing, especially to arrest the growth of *A. maculatus*, flew in the face of basic tea planting methods. But the biggest paradox, and obstacle, in his estimation was the practice of rice cultivation in the tea gardens that effectively canceled out his previous findings: "if rice-growing need not be considered then the situation can easily be dealt with by draining and oiling combined, flooding, or by jungle-growing combined with draining."[75] He struck at the proverbial hornet's nest by suggesting an inversely proportional relationship between paddy cultivation and species-specific malarial control:

rice cultivation and only perhaps a mitigation of the malarial prevalence, or the rice given up and a non-malarious labour force.[76]

Strickland's scorn for the "irresolute squad of managers" and their "fetish" of doling out cultivable land to laborers, points to a larger, and intractable relationship between disease prevention, health, and profiteering in the Assam estates. But how so? While we discuss specifics of this debate in the next section, suffice it to note for now that geographical distance, middlemen involvement, and economic competition meant that labor recruitment to Assam was a costly affair. Especially after 1882 therefore, planters resorted to penal provisions of existing labor laws to overwork workers while keeping wages to a bare minimum.[77] In addition, the desire for a readily available pool of recruits, even after the expiry of terms of contract, meant that a pattern of permanent settlement in and around the tea gardens was preferred to unfettered cyclical migration. Of course, the fact that laborers could come up to Assam under a concurrent scheme, without indenture and thereby more "free" and remuneratively demanding, led to intra-plantation rivalry and labor trafficking.[78] Under these circumstances, planters

---

[75] Ibid., p. 25.   [76] Ibid., p. 25.
[77] In fact, Behal and Mohapatra argue that despite a mortality rate of 6.4 percent in Act laborers in 1889, only "five per cent of total working days in Assam were granted as leave of absence due to illness." They also suggest that the penal contract system "militated against a normal rate of reproduction by the labour force ... averaging only 86 per 1000, compared with an average of 127 births per 1000 women in the non-tea garden population of Assam" during the period 1880–1901, in Rana P. Behal and Prabhu P. Mohapatra, "Tea and Money Versus Human Life: The Rise and Fall of the Indenture System in the Assam Tea Plantations 1840–1908," p. 160.
[78] The two labor laws referred to here are the Act I of 1882, and the Workmen's Breach of Contract Act, XIII of 1859; see the next section for details.

perceived cultivable plots of land for rice and other vegetables as an attractive sop, both for intending recruits and to induce them to stay on.[79] Furthermore, this was a measure that allegedly took care of attendant problems – subsistence earnings, malnourishment, and an emotional reminder of village life back home. Ironically therefore, calls to proactive malarial control measures that cut into these logics fell on deaf ears even as mortality figures continued to rise, nearing 12.5 percent during the first decades of the twentieth century.[80] Dr. G. C. Ramsay, long-time medical officer in Cachar and later president of the Assam branch of the British Medical Association, was even more forthright in drawing parallels between human mismanagement and malaria in Assam. Addressing the Seventh Congress of the Far Eastern Association of Tropical Medicine and Hygiene in Calcutta in December 1927, Ramsay cautions:

the balance of nature has for many years been upset in many Assam tea estates when virgin jungle was felled and many artificial breeding areas created. Our obvious duty, therefore, is to assist nature to regain the balance by methods acceptable to civilized and unsalted mankind in the unnatural environment which he has created for himself. Truly, the cost of appropriate anti-malaria measures is trivial compared with the economic loss caused by this disease apart from the terrible wastage of life, ill-health and misery it creates.[81]

---

[79] The idea of tea-labor-turned-part-time-agriculturalist was not theoretical alone. For instance, in 1888, out of the 4464 contracts expired in the district of Lakhimpur in Assam, 788 settled down as cultivators while in Nowgong, the total acreage of land by ex-garden laborer was 1224. It was also reported that in Sylhet, "nearly [every coolie] has a cow, and the managers encourage the purchase of cattle as a means of attracting the coolies to the garden; some coolies own bullocks which they hire to the garden and to the villagers, and advance money to cultivators; that about 999.65 acres of land were granted to coolies up to the 31st December 1888, and that most of them who hold land continue to do garden work," vide *Report on Labour Immigration into Assam for the Year 1888* (Shillong: Assam Secretariat Press, 1889), especially chapter I. This practice was reiterated in the unpublished 1870 journal of the Scottish surgeon-planter David Foulis, albeit with a liberal dose of idyllic flourish: "here and there where the coolies is an object of interest and care to the planter we find extensive vegetable garden from which [they] can have the chance of obtaining wholesome additions to their scanty fare ... every hut ought to have its fruit tree in front, guava, jack or papaya, under the shade of which the tired *kodally walla* (or spade-worker) may smoke his hookah in the evening," in *The Tea Assistant in Cachar*, MS 9659, National Library of Scotland Manuscript Collection, p. 11; also, T. C. Crawford, *Handbook of Castes and Tribes Employed on Tea Estates of North-East India* (Calcutta: Indian Tea Association, 1924); Strickland's critique has to be read within these, and similar, arguments.

[80] Ralph Shlomowitz and Lance Brennan, "Mortality and migrant labour in Assam, 1865–1921," p. 105.

[81] G. C. Ramsay, "The Factors Which Determine the Varying Degrees of Malarial Incidence in Assam Tea Estates and the Fundamental Principles Governing Mosquito Control of Malaria in Assam," *Transactions of The Royal Society of Tropical Medicine and*

Figure 4.3 "Capt. Robertson and Hospital Assistant throwing quinine into the mouths of loaded coolies." Sadiya, 1911–12 © The British Library Board, Photo 1083/34(163)

In his official report of 1929, Strickland too takes note of the economic loss at the "altar of ignorance" and writes with incredulity that estate medical officers had not been consulted regularly in matters of site selection.[82] In conclusion, he suggests emphatically that quinine prophylaxis alone will be insufficient in rooting out malaria from the Assam gardens, a wisdom shared by Sir Malcolm Watson earlier in 1924.[83]

These debates about cholera, kala-azar, and malaria etiology foreground an important idea, namely that ill health in the Assam estates exceeded medical logics of bodily disorder. While scientific

Hygiene, Vol. XXIII, No. 5 (March 1930): 511–518; see also his Obituary in *British Medical Journal*, June 6, 1959, 1(5135): 1478.

[82] See C. Strickland and K. L. Chowdhury, *Abridged Report on Malaria in the Assam Tea Gardens: With Pictures, Tables and Charts* (Calcutta: Indian Tea Association, 1929), pp. 101–102.

[83] See Sir Malcolm Watson, "Observations on Malaria Control, With Special Reference to the Assam Tea Gardens, and Some Remarks on Mian Mir, Lahore Cantonment," *Transactions of the Royal Society of Tropical Medicine and Hygiene*, Vol. XVIII, No. 4 (October 23, 1924): 147–154; on the Mian Mir experiment, see also Nandini Bhattacharya, "The Logic of Location: Malaria Research in Colonial India, Darjeeling, and Duars, 1900–30."

wisdom and sanitarians blamed multiple factors – germs, "bad batches," miasmas, hostile disease environment, species endemicity, non-immunity, and imported epidemics – as routes of disease, the root of the problem was not so well defined. On the one hand, attitudes to health in the Assam plantations depended on the political economy of profits and returns as discussed. On the other hand, the domain of law that regulated labor immigration and sanitary welfare, especially after Act I of 1882, created further discursive and enumerative understandings of fitness and well-being.

Laws of Health: Paradox as Problem

> There can be no doubt that Act I of 1889 and the rules framed thereunder have most materially decreased the mortality among immigrants.[84]

To historicize the above views of a provincial administrator in 1891 is to wade into the dense labor laws controlling immigration to the Assam estates. While details of these legislations do not concern us here, the social history of health in these plantations cannot be understood outside of, or dissociated from, its legal context. Beginning with Act III of 1863, Act VI of 1865, and Act VII of 1873, the entangled histories of law, life, and labor in the Assam plantations is hard to separate.[85] The 1865 Act effectively introduced the penal contract indenture system to Assam, nine-hour workday, and overriding powers to the management to arrest absconding workers without warrant. The balance of payment game was of course a two-way process: difficulties in transport, dearth of labor, and the unchecked growth of fraudulent middlemen exponentially increased the per head cost of recruitment for planters; insanitary conditions, low wages, malnourishment, and estate servitude led to high mortality, and persistent bouts of ill health among workers.[86] Pathogens, locality, or poor physique could only be blamed as much for causing sickness and death in these gardens.

---

[84] See Assam Secretariat Proceedings, Rev-B, No. 275/298, October 1891, p. 5, ASA.
[85] Also see Nitin Varma, "Coolie Acts and the Acting Coolies: Coolie, Planter and State in the Late Nineteenth and Early Twentieth Century Colonial Tea Plantations of Assam," *Social Scientist* Vol. 33, No. 5/6 (May–June 2005): 49–72; and, Varma, *Producing Tea Coolies? Work, Life and Protest in the Colonial Tea Plantations of Assam, 1830s–1920s*, Unpublished D.Phil. Dissertation (Berlin: Humboldt University, 2011).
[86] By one estimate, out of the 85,000 laborers brought into Assam between 1863 and 1866, 35,000 were reported to have either died or deserted, quoted in Rana P. Behal and Prabhu P. Mohapatra, "Tea and Money Versus Human Life: The Rise and Fall of the Indenture System in the Assam Tea Plantations 1840–1908," *Journal of Peasant Studies*, 1992, Vol. 19, No. 3: 147.

Interestingly, since 1864, a concurrent scheme to recruit labor outside of these acts, namely Act XIII of 1859 was also extended to Assam. Though mostly used to recruit local and contract-expired workers, planters preferred the penal provisions of the other legislations in order to keep wages to a bare minimum while maximizing productivity as stated. More importantly, after 1865, two classificatory schemes were introduced – "Act Labor" for those immigrating under these legislations and "Non-Act" – for workers re-engaging in service after contract expiry and therefore theoretically "free"; for laborers recruiting under the 1859 act and thereby outside indenture; and for non-productive and infirm dependents of Act-Labor. These enumerative logics are also important for our history of health, as much of the recorded assessment of disease in these estates (whether by sanitarians or estate medical men) depended on the identifiable, and therefore the Act population into account. While statistical veracity of mortality figures, and indices of health in the Assam plantations remain contested to this day, the Non-Act enumerations are even more problematic. Despite their yearly "appearance" in immigration report, Non-Act-Labor was theoretically outside the administrative control of planters and heavily under- or un-reported in mortality figures.[87] For *Vibrio cholerae*, *Leishmania donovani*, and the *Anopheles*, of course, legal status had little meaning.

The international tea market, especially in terms of prices, took a turn for the worse after 1878 and Assam planters, already saddled with vast swathes of plantation land sought greater returns for their investment. Out of all the factors of production, labor was the most malleable and the industry argued that its "future hinged on the maintenance of an adequate supply of coolie labour at a cost calculated to leave a fair margin of profit."[88] The government in council buckled under intense lobbying with Act I of 1882, the legislative mainstay for the Assam plantations till the first decades of the twentieth century. The 1882 act increased the maximum terms of contract to five years, deregulated recruitment (though still under government supervision), fixed minimum wages and retained the penal provisions of law. Major sanitary stipulations for the protection of labor health were instituted, though these were to remain paper tigers for the most part, as we shall see. Ironically, the idea of deregulation proved to be a major flaw and the number of unscrupulous and unlicensed recruiters, contractors, and middlemen shot up. Also, very few workers renewed their contract for the full five years preferring

---

[87] See Ralph Shlomowitz and Lance Brennan, "Mortality and Migrant Labour in Assam, 1865–1921," pp. 92–94.
[88] Behal and Mohapatra, "Tea and Money Versus Human Life," p. 147.

the flexibility of the 1859 act instead. Desertions and general exodus to "healthier" gardens were not infrequent throughout this period. By the turn of the century, both economic and political climate turned unfavorable for planters, and amidst reports of plantation violence and "return to slavery in the British dominion,"[89] the government was forced to retract some of the penal provisions of earlier labor laws. Interspersed with official labor enquiry commission reports, the government rescinded unregulated recruitment and penal provisions of earlier labor laws. Act VI of 1901 and XI of 1908 signaled the beginning of the end of this process. By 1926, the entire indentured apparatus, including Act XIII of 1859 was a dismantled ruin.

The tortuous history of these acts show that legislating health was an ongoing process in the Assam plantations. Of course, with burgeoning costs and demands for more working hands, hygiene, nutrition, sanitation and welfare came towards the end of management priorities. But fitness and productivity were directly proportional. As planters treaded a tenuous line between sanitary investment and profit margins, law became a cause célèbre negotiating and setting parameters of sickness and well-being in Assam. But the problem was in the paradox, for if more laws meant better health, the situation in these plantations proved otherwise. On their part, planters blamed legal stringency and loopholes – along with germs, miasmas, and non-immunity – as causing death in the province. As a theoretical barometer, an etiology as I argue here, law was intimately connected to health in these plantations of eastern India.

In their report of 1868, the commissioners appointed to examine the state and prospects of tea cultivation in Assam candidly remarked on this relationship.[90] Especially critical of the 1863 act for lacking regulatory teeth and clarity, they suggested that it had failed to provide for medical inspection at labor disembarkation depots, restrict overcrowding aboard steamers, and check unlicensed contractors from whisking away recruits bypassing its provisions.

---

[89] See Dwarkanath Ganguly, *Slavery in British Dominion*, edited Siris Kumar Kunda (Calcutta: Jijnasa Publications, 1972); Sir J. H. S. Cotton, *Indian and Home Memories* (London: T. Fisher Unwin, 1911); Mrs. Emma Williams, "Letter regarding abuses on the tea plantations of Assam," IOR/L/PJ/6/749, March 24, 1906, British Library London; Report from Aborigines Protection Society on "Treatment of tea labourers in Assam," IOR/L/PJ/6/193, January 17, 1887; Revered C. Dowding, "Letters and pamphlets on the illegal arrest of run-away tea-garden coolies in Assam," IOR/L/PJ/6/832, October 22, 1907, and the numerous House of Commons Parliamentary papers on the topic.

[90] See *Report of the Commissioners Appointed to Enquire into the State and Prospects of Tea Cultivation in Assam, Cachar and Sylhet* (Calcutta: Calcutta Central Press Company Ltd., 1868).

Without getting into the medical discourse of cholera etiology, the commissioners argued that "there is little doubt that the germs of it have been imbibed in the depot."[91] Drawing attention to the high mortality rates of Assam immigrants vis-à-vis other colonial émigrés, the report points out that the 1863 act had been followed more in the breach than in practice. In their assessment:

> The results of the working of Act III [of 1863] appear to us to have been far from satisfactory. The death-rates, both in the depots and on the voyage, have been extremely high. What these rates were before the Act came into operation, we have been unable to learn, but the mortality attendant on the emigration which has taken place since it came into force has been very great.[92]

Medical inspection of labor was a thorny issue, for strict parameters of fitness and planter demands for working hands were often incommensurable. Moreover, the government was unclear if its authority extended to deciding questions of bodily ability and health on behalf of the employer, or planter in this case. As it is, the 1863 act had left this debate wide open. Thus, in a letter dated October 17, 1864, its stand was clarified: "the fitness of the labourer for work is a matter which it is not the duty of the Government Medical Inspector to decide. If a laborer decides to go and is not, in the opinion of the Medical Inspector, unfit for the voyage, that officer should offer no hindrance to his going."[93] With mortality figures mounting, these tautological arguments were both unconvincing and ineffectual. Act VI of 1865 amended some of these indecisions by stipulating hospitals for every estate and a labor protector. For the 1868 committee, the function of law went further:

> The object of special legislation ... has been to afford protection, so far as possible, both to the laborer and to the employer; but if men are allowed to proceed who are unfit to labor, and who, from natural weakness of constitution, are likely to succumb early to the effects of a new and malarious climate, the interests of both parties are sacrificed at the very outset. If the medical inspection be rendered more strict, the field of recruitment will certainly be limited; but the amount of real labour placed at the planter's disposal will be little effected, while he will be saved the fruitless expense, which so many have been put to, of bringing up labourers who are of no use.[94]

Here, in this version, law fulfilled several epidemiological functions. The medical term is apt as, for the committee at least, law could both *cause* (as with the 1863 act) and *control* (as with the 1865 act) mortality in the Assam gardens. More importantly, law was the remedy – for laborers, by stopping unhealthy souls from proceeding onwards and for

---

[91] Ibid., p. 47.   [92] Ibid., pp. 35–36.   [93] Quoted in ibid., p. 41.   [94] Ibid., p. 42.

employers – by providing them with the best, albeit restricted "crop" of able-bodied men. It was an argument to resolve the paradox of health, morbidity, and productivity all at once. In the aftermath of the tea mania, these intertwined medico-legal logics did not seem out of place. But sanitary science was not an abstraction, and the 1868 committee realized that its argument could only be sustained if the protector was also a medical man and therefore able to help laborers "in important matters affecting health, in which experience has shown that he requires protection most."[95] Ground realities were of course quite different,[96] and the inquiry admitted that a labor inspector, rather than protector, was more practical under the circumstances. For the plantation management, labor legislations proved to be a necessary evil. If it facilitated immigration under a tough labor market, excessive interference in matters of health and sanitary welfare eroded authority and ate into profits. For once, even the lieutenant governor of Bengal sided with planters and argued that it was "a most intolerable arrangement to compel him to employ according to law any Medical Officer whom the Government may choose to select for him."[97] The tug-of-war gained more steam over the next four decades.

Laissez-faire politics aside, legislating health in these plantations was ideologically necessary and materially expedient. With unabated deaths in the Assam plantations, Act I of 1882 envisioned a more proactive role in defining and regulating worker well-being. Despite its stringent penal contract clause, the 1882 act provided a veritable prescription list of sanitary measures that contractors, recruiters, embarkation agents and planters were expected to follow. Amidst the scientific indecision on disease etiology, the logic of law returned with even greater urgency with its mandate to control sickness and mortality. Thus, the inspector was now empowered to legislate on labor housing, water supply, sanitary arrangements, estate hospitals, diet, ration, ability to work, and compensation due

---

[95] Ibid., pp. 76–77.
[96] An overreaching protector could also induce crisis of authority among planters. In one such instance, Mr. A. P. Sandeman, planter in the Dibrugarh district of upper Assam noted in his testimony of January 1, 1868 that the presiding Protector, Mr. Marshall had substantially eroded his standing among laborers by assuming the role of de-facto manager. To make matters worse, "in the case of pregnancy," Mr. Marshall "ruled that a woman should have leave for both one month before and one month after her confinement at full pay of Rs. 4 a month"; a similar complaint was made by Mr. J. M. Wood, Manager of the Nagagooli Plantation of the Upper Assam Tea Company in his letter dated January 3, 1868, both quoted in *Report of the Commissioners Appointed to Enquire into the State and Prospects of Tea Cultivation in Assam, Cachar and Sylhet*, p. xxxiii and p. xl, respectively.
[97] Vide, Bengal Government Papers, Emigration, File No. 303/5999, p. 11, July 1869, ASA.

Death in the Fields 125

to illness.[98] Additionally, under sections 143 and 144 of the 1882 act, the local government was empowered to institute more specific laws to do with labor transportation en route to Assam. Accordingly, contractors and employers were henceforth asked to see that medical officers be provided at depots, vaccinations of emigrants completed, and infected clothing destroyed. Also, steamers needed to have adequate ventilation, four water tanks, six taps for the "purest drinking water obtainable," troughs for washing, arrangements for bathing, two sets of latrines for men and women, lanterns for safety, and fire-buckets in the event of an emergency.[99] These logics did not, however, rule out "a particular class of labourers" unfit for plantation work from its legal vocabulary.

In addition to legal instructions on sanitation, numerical jousting also instituted parameters of health and sickness during this period. Thus, any estate exceeding an annual mortality rate of 7 percent of the total population was henceforth blacklisted and classed "unhealthy."[100] In 1889, in a bid to "strengthen control" over persistent mortality, it was decided that 7 percent would now apply separately – for Act- and Non-Act-Labor whereas it had only been applied cumulatively in the past. The result of this statistical maneuver spoke for itself (see Table 4.1 below).

The paradox was obvious; more laws had not only failed to curb unhealthiness, but had also created unreliable indicators of well-being in the first place. Consider the wide variation in the number of "unhealthy" gardens using the two methods between 1889 and 1895. In any case, gathering data was in itself a tricky problem and often highlighted the administrative limits of the colonial government. Reminders and urgent requests for mortality figures from planters and district surgeons frequently fell on deaf ears.[101]

---

[98] See "The Assam Labour and Emigration Act I of 1882," in *The Assam Code: Containing the Bengal Regulations, Local Acts of the Governor General in Council, Regulations Made Under the Government of India Act, 1870, and Acts of the Lieutenant-Governor of Bengal in Council, in Force in Assam, and Lists of the Enactments which have been Notified for Scheduled Districts in Assam under the Scheduled Districts Act* (Calcutta: Office of the Superintendent of Government Printing, 1897), pp. 173–174.
[99] See *Rules Under the Inland Emigration Act I of 1882* (Calcutta: The Bengal Secretariat Press, 1884), especially chapters I, II and IV.
[100] There is no historical clarity on how the number seven was arrived at in the first place. Rajani Kanta Das argues: "these figures were more or less arbitrary, but they were fixed with a view to excluding gardens with a small force of workers from too easily falling into this category and at the same time to including those gardens where death-rates had been large enough to justify such declaration," in *Plantation Labour in India*, p. 105.
[101] Vide, Assam Secretariat Proceedings, Revenue-A, "Mortality on tea gardens in Assam," No. 55/73, May 1898, ASA.

Table 4.1 Statement showing mortality on tea estates and number of unhealthy estates, Appendix H, Assam Secretariat Proceedings, Emigration-A, File No. 229/4189 R, September 1896, ASA

Year	Number of new immigrants imported during the year	Average labor force	Number of unhealthy gardens calculated By old method in force before 1889	Number of unhealthy gardens calculated By new method adopted in 1889	Total tea-garden death-rate (percent)
1876	34,283	124,323	...	...	51.1
1877	31,897	146,513	...	...	50.9
1878	43,061	172,569	...	...	66.1
1879	24,712	184,935	...	...	50.9
1880	15,913	188,497	48	...	35.2
1881	17,116	194,182	60	...	31.7
1882	22,559	200,099	57	...	37.8
1883	32,138	279,867	73	...	41.3
1884	45,511	267,855	93	...	43.2
1885	29,398	289,574	42	...	36.8
1886	30,894	301,349	64	...	39.8
1887	36,463	320,408	49	...	36.2
1888	46,293	347,371	65	...	39.8
1889	55,658	386,532	69	118	41.8
1890	36,080	406,089	28	87	34.3
1891	49,908	429,148	30	89	36.8
1892	56,050	457,717	55	143	41.9
1893	50,675	479,743	14	75	33.2
1894	46,530	494,336	21	86	31.9
1895	72,837	526,833	11	70	33.6

In a candid disclosure, the government was forced to admit in 1890 that "deaths have been imperfectly classified ... due to the non-receipt of returns from managers and the dilatoriness on the part of the Civil Surgeon to inspect."[102] History repeated itself once again, and an official 1893 inquiry into the working of the 1882 act revealed widespread discrepancy between policy and practice. Legislating health had serious operational, and economic limitations. It soon dawned on the government that race prejudice had forestalled bathing provisions in labor depots, as "emigrants of the type known as 'jungly' [were] not accustomed to daily washing."[103]

---

[102] Vide, Assam Secretariat Proceedings, Revenue-B, No. 275/298, October 1891, ASA.
[103] Vide, Assam Secretariat Proceedings, Revenue-B, No. 462/483, December 1893, ASA.

In addition, vaccinations of adults and dependents were rarely carried out, "burial of excreta" never enforced, and latrines inadequate. Surgeon-Major J. Mullane argued that law might even be indirectly responsible for worker malnutrition in the Assam estates. He testified that rice being more expensive in the open market, the stipulation to feed workers at contract rates proved unprofitable for employers who therefore kept its supply as low as possible.[104] Similarly, a large number of ancylostomiasis deaths in 1888 were attributed to making "coolies palpably suffering from it ... to go on working, until it develops into an incurable stage."[105] A memorial by the Indian Association to the GOI that year argued that planters had used the penal provisions of the 1882 with impunity to overwork laborers, especially women. High infant death rates (around 44 per thousand in 1884) and low birth rates (32.7 per thousand the same year) were linked to the impossibility of maternal care, neglect, lack of leave time, and preference for abortion to infantile servitude in the gardens. The memorial begged the government to inquire into aspects "of the law that makes such things possible."[106]

Planters of course had long argued that interference itself was at the root of the problem, and that Act I of 1882 was "too inquisitorial."[107] Blaming labor non-immunity (or "bad batches") for rising mortality had been a favorite trope for managers and provincial civil surgeons alike. Sometimes, law attempted to override this incommensurable logic. In a letter dated September 30, 1896, the Officiating Secretary to the Chief Commissioner of Assam clarified its position to GOI:

It should be explained that the importation of bad batches is, under no circumstances, accepted by the Chief Commissioner as sufficient excuse for excessive mortality on an estate; but in cases in which such importations have taken place, with the result of high mortality, and the employer is found to have neglected the precaution of having his labourers medically examined before placing them under contract, the Chief Commissioner has the less hesitation in putting in force against him the coercive provisions of the Act.[108]

---

[104] See "Report by J. Mullane, MD, Surgeon-Major and Civil Surgeon," dated April 16, 1893 in ibid., ASA.
[105] Quoted in Rana P. Behal and Prabhu P. Mohapatra, "Tea and Money Versus Human Life: The Rise and Fall of the Indenture System in the Assam Tea Plantations 1840–1908," pp. 159–160.
[106] Vide, Secretary to the Indian Association to the Secretary to the Government of India, April 12, 1888, IOR/L/PJ/6/257, Asian and African Studies, British Library, London.
[107] See *The Indian Planters' Gazette and Sporting News*, July 6, 1886, p. 1.
[108] Vide, Assam Secretariat Proceedings, Emigration-A, File No. 229/4189R, September 1896, pp. 12–14, ASA.

Table 4.2 Statement of adult death rates in a twenty-year period; *vide, Annual Reports on Labour Immigration into Assam for these years* (Shillong: Assam Secretariat Press)

Year	Death Rate per Thousand	
	Act	Non-Act
1882	67.9	29.3
1883	61.5	30.0
1884	58.9	29.5
1885	51.5	26.5
1886	58.9	29.5
1887	57.2	25.8
1888	62.9	30.2
1889	63.5	35.6
1890	55.6	28.8
1891	49.8	34.1
1892	64.3	37.7
1893	51.7	28.5
1894	48.9	30.3
1895	52.8	31.2
1896	45.7	30.8
1897	56.7	41.9
1898	47.0	31.1
1899	43.6	26.3
1900	43.5	26.2
1901	37.2	25.2
1902	40.3	25.2

With the strict sanitary regulations of the 1882 act, and persistent fears of a stricter amendment, planters targeted law as their whipping post. They even ridiculed statistics of unhealthy gardens and argued that similar parameters of well-being would "sweep away whole villages" in mainland India.[109] Most employers throughout this period, however, recognized that the 1882 act, especially with its penal provisions, gave them "exceptional powers,"[110] sanitary diktats notwithstanding. In the closing decades of the nineteenth century, renewed attempts were made to regulate the health and welfare of tea workers headed to Assam. The Inland

[109] *The Indian Planters' Gazette and Sporting News*, January 12, 1886, p. 26.
[110] Ibid., July 6, 1886, p. 2.

Emigrants' Health Act 1889 was extended to the province, and rehearsed earlier norms for recruiting agents and labor depots. Though this law attempted to standardize sanitary provisions, food quality and water supply for laborers *en route*, the agents could now be fined for defaulting and his "money recovered together with a simple interest of 6 percent yearly."[111] Mortality numbers, however, continued to tell a different story (see Table 4.2 above).

As the new century dawned on these gardens of Empire, the looming presence of mortality and ill health refused to go away. Despite scientific innovations, special legislations, medical breakthroughs and vector identification, cholera, kala-azar, malaria, dysentery, and anemia continued to ravage the Assam estates. Act VI of 1901 made another attempt to regulate health, and strengthened the jurisdictions and power of the Medical Inspector while increasing the minimum wage for both men and women workers.[112] As it is, competition from other industries, and a volatile political and economic climate, diluted planter hold over labor during the first two decades of the twentieth century. As the penal clause came under fire, the colonial state argued more and more in favor of "freer" immigration as elixir for Assam's persistent, and peculiar labor problem. It was felt that health, too, would be better off under laissez-faire than law. The 1906 labor commission recommended: "the aim now is to keep the labourer on the garden by making life attractive to him and not by force of law … [we] are inclined to believe that a policy of less medicine and more leave off work would show better results as regards the health of labourers."[113] The paradox had come full circle in the Assam plantations.

Health, Disease, and the Labor Body

Away from the administrative gaze of the Presidency towns and municipal areas, disease etiology, prophylactic policy and parameters of health in the Assam tea plantations emerged from a strategic and uneasy relationship between the colonial state, medical opinion, and planter interests. This chapter has shown that, as

---

[111] Vide, Act I of 1889, Passed by the Lieutenant-Governor of Bengal in Council, May 7, 1889, p. 4, IOR/L/PJ/6/257, Asian and African Studies, British Library, London.

[112] Vide, The Assam Labour and Emigration Act, VI of 1901, especially chapters II, III and IV in F. G. Wigley, *The Eastern Bengal and Assam Code: Containing the Regulations and Local Acts in Force in the Province of Eastern Bengal and Assam*, Vol. I (Calcutta: Superintendent of Government Printing, 1907), pp. 527–594.

[113] *Report of the Assam Labour Enquiry Committee, 1906* (Calcutta: Superintendent of Government Printing, 1907), pp. 71–97.

far as the "economics of Eden"[114] was concerned, the body of the Indian laborer was primarily for productive work. His/her accessibility to, and availability for, the ambitions of Western medicine – ideological differences notwithstanding – was conditioned by the imperatives of commodity capital and exigencies of plantation life. Our discussions of cholera, kala-azar, and malaria investigations in Assam amply highlight this tension between epidemiological theory and practice.

Disease etiology, identification, and prophylaxis also varied depending on whether the economic, legal, or natural body was being called on. Thus, Surgeon DeRenzy wore his staunch contagionist hat while dealing with Assam tea labor cholera and swiftly sacrificed it when it came to his own company's unhealthiness. Similarly, Haffkine's vaccination experiments were both facilitated by the aggregate pathological body of the Assam plantation workers, and constrained by its economic rhythms. Strickland's prognosis of planter policy, site selection, and the anopheles as the unholy trinity of Assam's malaria problem also invoked these intertwined logics. Law, too, often spoke in a forked tongue: Act I of 1882 thus concurrently blamed hasty recruitment, insanitary work conditions, and poor "class of labour" for death and mortality in the tea gardens of eastern India. In other words, disease – as concept and condition – mediated institutional agendas (CSTM, ITA), individual opinion (Giles, Rogers, Strickland et al.), special legislations (1863, 1865, 1873, 1882, 1901), commodity markets, and bodily culpability ("unsuitable class," "non-immune race," imported epidemicity) in questions of cause and cure. Of course, etymologically, the definition of disease in Assam was in itself slippery. Consider that malaria was subsumed under "jungle fever" till 1892, that ancylostomiasis and kala-azar were undifferentiated. As Charles Rosenberg argues: "it is fair to say that ... a disease does not exist as a social phenomenon until we agree it does – until it is named."[115]

The legal, economic, and ecological interpretations of the body are not hermetically sealed. In fact, the scientific, functional, and regulatory transcripts of health and disease in Assam are inseparable from each other. All this is not to suggest, however, that Assam was pathologically unique, or that it formed an exceptional disease

---

[114] I borrow this phrase from Richard Drayton; see his *Nature's Government: Science, Imperial Britain, and the "Improvement" of the World* (New Haven, CT: Yale University Press, 2000), p. 80.

[115] See Charles E. Rosenberg, "Framing Disease: Illness, Society, and History," in Charles Rosenberg and Janet Golden, eds., *Framing Disease: Studies in Cultural History* (New Brunswick, NJ: Rutgers University Press, 1992), p. xiii.

Death in the Fields 131

zone.[116] Rather, I have argued that ideas of mortality, death, and well-being exceeded – indeed had to exceed – instrumental logics of scientific objectivity, imperial sanitary policy, vector identification, and preventive cure in these estates. It was shaped by the expedient exegesis of medical knowledge, law, commercial interests, and idioms of corporeality.

The tea industry's production demands alongside unabated labor mortality rates created an unprecedented challenge for the colonial government. It is fair to say that in the tea gardens of eastern India, the colonizing force of Western medicine had to adapt to, if not be upstaged by, the demands of capital. As it is, authority and power were zealously guarded privileges in the Assam estates, and planters reluctantly shared it with sanitarians, investigators, and medical men. Unlike jails, therefore, the colonial state only had mediated access to tea workers in matters of health and disease control. But the meaning of labor well-being went one step further in the Assam plantations. Along with the pathological and economic, law infused an additional layer of discursive (and enumerative) understanding of the body, health, and morbidity in these plantations.

The special legislations aiding immigration to the province were inherently contradictory and Janus-faced. While it pandered to planter requests for "freer" recruitment and the penal contract, these laws simultaneously attempted to regulate labor health and sanitary welfare.[117] But this was a paradoxical vision. As shown, strict parameters of bodily fitness and the demands of plantation labor were incommensurable, or even unprofitable for the most part. In its turn, the terms of penal indenture, and recruitment loopholes were themselves the cause célèbre for ill

---

[116] To be sure, there is a strain of medico-historical discourse that apostrophized Assam (and its tea estates) with its own disease identity, and "primitive" pathologies. Kala-azar was variably referred to as "Assam fever," a "disease of the plantations," or "coolie fever" in colonial accounts and medical reports; its etiology traced to "backward" norms of hygiene and bodily behavior; see Bodhisattva Kar, "The Assam Fever: Identities of a Disease and Diseases of an Identity," in Debraj Bhattacharya, ed., *Of Matters Modern: The Experience of Modernity in Colonial and Post-colonial South Asia* (Calcutta: Seagull, 2008), pp. 78–125; for a parallel discussion on the historical ecology of malaria, hookworm, and pellagra in the American South, see Todd L. Savitt and James Harvey Young, eds., *Disease and Distinctiveness in the American South* (Knoxville, TN: The University of Tennessee Press, 1988); also see Ronald L. Numbers and Todd L. Savitt, eds., *Science and Medicine in the Old South* (Baton Rouge, LA and London: Louisiana State University Press, 1989).

[117] To be sure, the contradictions inherent in these labor laws were part of its overall character, and extend beyond these bodily, indeed medical ramifications. As discussed briefly in the beginning of section four, Act I of 1882 in trying to deregulate recruitment only occasioned abuses in its name. In highlighting the incommensurability of "protection" and "productivity," this chapter has examined only one subset of the labor law paradox – namely in terms of health and morbidity.

health, death, and disease en route to and on the plantations of Assam. Conversely, planters rued the excessive and unrealistic sanitary goals of these laws as "causing" sickness, and argued for more freedom as the panacea for better health. Often, the colonial government sided with these warped medico-legal logics.

Managing health and disease in these estates rested on a functional balance between economic, legal, and medical viability; its history has to be therefore understood alongside the structure of the plantation industry and its peculiar relations of production. Here, in the "garden of the Lord,"[118] ill health had multiple claimants: germs, miasmas, profiteering, and law.

---

[118] This phrase was used by Nathan Brown, the American Baptist missionary on his travels to upper Assam in 1836, quoted in H. K. Barpujari, ed., *The American Missionaries and North-East India, 1836–1900* (Guwahati: Spectrum, 1986), pp. 7–8.

5   Conservation or Commerce?

> The existence of well-ordered and protected forest estates in the farthest and most out-of-the-way places demonstrates [...] completeness of control extending to the farthest verge of the State domain, which characterises a strong Government.[1]

The importance of forests for tea is obvious. Plantations needed trees for shade, moisture retention, and soil regeneration. But it is also necessary to fell them for their wood in building house-posts, factories, tea-boxes and for charcoal for the firing stage of tea manufacture. If these needs contradicted each other in terms of preservation and profits, the Forest Department (hereafter FD) added a third dimension to this commodity story. Formally established around 1864 with a mandate to protect and conserve empire's forest wealth for posterity, the FD found itself a late entrant in the agro-economic spoils of Assam. Having had preferential access to the province's vast wastelands – first through outright sale and later through liberal long-term leases – the tea enterprise outbid the FD's ambition to redefine property relations by a significant temporal margin. As the FD took over the newly separated commissioner's province in 1874, it realized that large tracts of land, often with valuable forest cover, had fallen into the hands of planters and speculators. Progeny of the tea boom period (1854–65), these lands were often "locked up" without bringing them under cultivation or the plough. Despite attempts to legally demarcate government forest areas – using Forest Acts VII of 1865, VII of 1878, and Assam Forest Regulation VII of 1891 – under its total or partial control, these "private" tea estates (and other private forests) remained mostly outside its purview. But there was another paradox to contend with. If the FD's legitimacy and authority stemmed from husbanding valuable forests against wanton use, the limited availability of such woods (teak, for instance) in Assam, unestablished markets, and sparse local

---

[1] B. H. Baden-Powell, "The Political Value of Forest Conservancy," *The Indian Forester*, Vol. II, No. 3, January 1877, p. 284.

demand meant that the tea industry and its requirements was a primary customer – and "creator" – of forest wealth and worth in the province.

But this is not an institutional history of forestry in the province,[2] nor do I look at the entire gamut of issues involved in such a story. Rather, the focus here is on the complementary, and conflicting, agendas of these two patrons of colonial resource management in Assam. I argue that the FD and the tea enterprise shared an uneasy, but expedient, co-dependency in nineteenth- and early twentieth-century eastern India. Indeed, the parallel existence of these two monopoly sectors – one directed at forest capital, and the other at crop capital – called both competition and collaboration into play. For if the FD needed the tea industry, the latter could not do without government help in resolving its insatiable appetite for land, cheap charcoal and wood for tea-boxes. One therefore needs to look beyond narratives of resource depletion[3] and conservation dilemmas[4] to grasp the double-bind of these two agencies in Assam.

This discussion also adds to the longstanding debate in Indian forest history regarding the relationship between colonial silvicultural ideologies and their consequences on the ground. A significant burden of this historiography has thus been to show that ideas of commercial forestry and resource management – originating with the birth of the forest department and its laws – ruptured centuries of indigenous agrarian practices, priorities and livelihoods in the subcontinent.[5] These shifts, in turn, led to sustained peasant protests and rebellions that continue to this day.[6] Studies have also shown the impact of British

---

[2] For an in-depth analysis of the many aspects of forestry in Assam, see Arupjyoti Saikia, *Forests and Ecological History of Assam, 1826–2000* (New Delhi: Oxford University Press, 2011).

[3] Richard P. Tucker, "The Depletion of India's Forests under British Imperialism: Planters, Foresters, and Peasants in Assam and Kerala," in Donald Worster, ed., *The Ends of the Earth: Perspectives on Modern Environmental History* (Cambridge: Cambridge University Press, 1989).

[4] Arupjyoti Saikia, "State, Peasants and Land Reclamation: The Predicament of Forest Conservation in Assam, 1850s–1980s," *Indian Economic and Social History Review* Vol. 45, No. 1 (2008): 77–114.

[5] For the western Himalayan region of Tehri Garhwal, see Ramachandra Guha, *The Unquiet Woods: Ecological Change and Peasant Resistance in the Himalaya* (Ranikhet: Permanent Black, rpt. 2013); also, Madhav Gadgil and Ramachandra Guha, *The Use and Abuse of Nature* (New Delhi: Oxford University Press, 2005); for a general history of peasant protests in Assam that also traces its origins to imperial forest policies, see Arupjyoti Saikia, *A Century of Protests: Peasant Politics in Assam Since 1900* (New Delhi: Routledge, 2014).

[6] See Amita Baviskar, *In the Belly of the River: Tribal Conflicts over Development in the Narmada Valley*, second ed. (New Delhi: Oxford University Press, 2005); Nandini Sundar, *Subalterns and Sovereigns: An Anthropological History of Bastar (1854–2006)*, second ed. (New Delhi: Oxford University Press, 2008); Vandana Shiva, *Staying Alive: Women, Ecology, and Development* (New Delhi: Kali for Women, 1988), and

ideas of forest conservation on land ownership, shifting cultivation, and game management.[7] Still others have argued that the logics of colonial State-making and governance – as reflected through these forestry paradigms – was not a unitary imposition from above. Rather, these idioms of rule emerged in conflict with, and sometimes as a result of being confounded by, regional economies and their people, flora, and fauna.[8] To this list are "prosopographical" approaches and theoretical revisionists who challenge the purported "colonial" coming-of-age of these forest ideas and policies.[9] In this strand, colonial silviculture was but an extension – albeit an important one – of its Continental, especially German, roots. It calls for looking at the inter-imperial connections of the rise and spread of these novel sylvan policies and the individuals, institutions, and intellectual legacies that gave it shape and direction.

Amidst these concerns, this chapter shows that not all landscapes were for the forest department's taking, either by force or law. Indeed, the early advent of tea in Assam short-changed the forest department's ability to transform rights of land ownership and forest resource management. Within this context, the workings of "rational" scientific forestry in the province had to go beyond the dual options of preservation and conservation. Finally, if indigenous struggles against these forest principles are largely absent here, it is not because they are unimportant. While some of that story has already been told,[10] the overlapping and obstructing interests of these two major resource stakeholders demonstrate that they had taken over much of Assam local interests – and lives – during the nineteenth and early twentieth century.

Alpa Shah, *In the Shadows of the State: Indigenous Politics, Environmentalism, and Insurgency in Jharkhand, India* (Durham, NC: Duke University Press, 2010) to name a few.

[7] See Mahesh Rangarajan, *Fencing the Forest: Conservation and Ecological Change in India's Central Provinces 1860–1914* (New Delhi: Oxford University Press, 1996); also, Mahesh Rangarajan and K Sivaramakrishnan, eds. *Shifting Ground: People, Mobility and Animals in India's Environmental Histories* (New Delhi: Oxford University Press, 2014); also see Joseph Sramek, "'Face Him Like a Briton': Tiger Hunting, Imperialism, and British Masculinity in Colonial India, 1800–1875," *Victorian Studies*, Vol. 48, No. 4 (Summer 2006): 659–680.

[8] See K. Sivaramakrishnan, *Modern Forests: Statemaking and Environmental Change in Colonial Eastern India* (Stanford, CA: Stanford University Press, 1999).

[9] See S. Ravi Rajan, *Modernizing Nature: Forestry and Imperial Eco-Development 1800–1950* (New Delhi: Orient Longman, rpt. 2008).

[10] See Saikia, *A Century of Protests*; Saikia, "State, peasants and land reclamation"; Amalendu Guha, *Planter Raj to Swaraj: Freedom Struggle and Electoral Politics in Assam 1826–1947* (New Delhi: ICHR, 1977), and Jayeeta Sharma, *Empire's Garden: Assam and the Making of India* (Durham, NC and London: Duke University Press, 2011); also see the discussion on this point in the Introduction.

Land and its Lords

Planters were ... the biggest [...] landowners in every rural district.[11]

Let us revisit the tortuous land tenure arrangements that consolidated the hold of the tea enterprise in the province. For these policies provide the agro-economic context within which the FD was later to stake its operational claim.[12]

The land tenure policy that set the tone for these plantations was rooted in paternalistic favoritism. Almost from the start, the colonial government went all out to placate the planter lobby in doling out land and resources for tea. While impoverishing the local agrarian sector, these land grants made successive overtures to the planting community with the hope of attracting British capital and speculators.

Though Assam came into British possession after 1826, it was the Tea Committee's enthusiasm about the crop's prospects that initiated systematic agrarian colonization in the region. At the time, Dr. Nathaniel Wallich, superintendent of the Calcutta Botanical Gardens urged Captain Francis Jenkins, Agent to the Governor-General of the North-East Frontier in 1836 that "frontier" lands under "tribal chiefs" and "rajahs" be placed under proper management for establishing tea estates. While details of the colonial takeover of "tribal" lands for tea do not concern us here, Jenkins is a noteworthy figure. He soon decreed that in order to attract British entrepreneurs to tea, it was necessary that land grants be distributed freely with: "no other condition than the payment of a fixed and unalterable rate of rent and absolutely unencumbered with any stipulation in regard to ryots or sub-tenants."[13] The Board of Revenue agreed to his terms, and a "liberal" scheme was thereafter introduced in August 1836.[14] Under it, three classes of land were earmarked – forest and high wasteland, high reed and grassland, and grasslands amidst already cultivated lands. The first, second, and third classes of land were to be held rent-free for five, ten, and twenty years

---

[11] See Amalendu Guha, "Assamese Agrarian Relations in the Later Nineteenth Century: Roots, Structure and Trends," *The Indian Economic and Social History Review*, Vol. XVII. No. 1, January–March 1980, p. 53.

[12] For a detailed study of land revenue systems in Assam, see B. H. Baden Powell *The Land Systems of British India*, Vol. III (Oxford: Clarendon Press, 1892) in addition to Guha's "Assamese Agrarian Relations."

[13] See Amalendu Guha, *Planter Raj to Swaraj*, p. 12. The term *ryot* (or *raiyat*) was used by the British to refer to a peasant cultivator; it also simply meant a peasant, or, during the Mughal period, to a subject of the empire.

[14] Cited in H. K. Barpujari, *Assam: In the Days of the Company 1826–1858* (Gauhati: Lawyer's Book Stall, 1963), p. 212; also see Muhammed Abu B. Siddique, *Evolution of Land Grants and Labour Policy of Government: The Growth of the Tea Industry in Assam 1834–1940* (New Delhi: South Asian Publishers, 1990), especially Chapter 2.

respectively. It was, however, required that one-quarter of the land be under cultivation by the end of the fifth year of the grant.[15] Designed to check wanton speculation, this stipulation empowered the government to take over lands that did not fulfill these terms.

Captain Jenkins soon expressed unhappiness with the 1836 rule. For him, it did not sufficiently empower *British* capitalists to take over lands for tea cultivation. Euphemisms aside, Jenkins rued the lack of exclusive European rights in land colonization and campaigned for a more favorable policy.[16] On March 6, 1838, the government acquiesced and a new set of orders was passed. Under Rule II, it was stipulated that "no grant for agricultural purposes on the advantageous terms which it is proposed to authorize be made of a less extent than one hundred English acres, nor of a greater extent than ten thousand English acres."[17] Furthermore, Rule III specified that: "no grant be made until the applicant shall satisfy the Collector that he is possessed of capital in money, grain, or agricultural stock, or implements, to the extent of Rs. 3 per acre." In a letter dated July 25, 1836, the Board of Revenue also included forest lands in these grants to be made available rent-free for twenty years.[18] Historically known as the Wasteland Rules of 1838, even a summary glance at the two sections referred to above indicates that European settlers were favored over the local Assamese. Despite these inducements, the Assam Company remained the only major player in tea production for most of the next fifteen years. Communication bottlenecks, labor recruitment challenges, and a slow and bumpy start to successful tea manufacture meant that between 1838 and 1850, only eighteen grants for a total area of 5,533 acres were handed out to the Assam Company. It dawned on the government that an even more robust and lucrative land grant system was called for.

On the eve of Governor-General Dalhousie's visit to Assam in 1853, Mr. A. J. M. Moffatt Mills, judge of the Sadar court in Calcutta, was deputed to report on the conditions in the province. He bluntly observed:

In a country like Assam, where there is a superabundance of land and a deficiency of labor, I strongly deprecate the granting of waste land to natives of the Province, except under peculiar circumstances: they have no capital, and their only resource

---

[15] Siddique, *Evolution of Land Grants and Labour Policy of Government*, p. 13.
[16] It was even argued in some circles that the 1836 Rules might induce the Assamese *ryots* to abandon their revenue-paying lands, take up wastelands, and return them at the end of the rent-free period, cited in Barpujari, *Assam: In the Days of the Company*, p. 213.
[17] See Appendix A, *Papers Regarding the Tea Industry in Bengal* (Calcutta: Bengal Secretariat Press, 1873), pp. 94–97.
[18] Vide letter no. 278, dated July 25, 1836, paragraph 3, in ibid., p. 94.

is to seduce other Ryots to settle in these grants, so that as much or even more becomes waste in one place than is reclaimed in the other.[19]

Mills' *Report* famously became the cause célèbre for another round of rule revision. In its wake, the rules of 1854 (known as the Old Assam Rules) were drawn up. In scope and vision, these rules expressly targeted the European tea capitalist to take up land for settlement and production. For instance, Rule III specified that: "no grants shall be for less than 500 acres of forest or grass waste, which will be granted on the same terms" while Rule V stated that: "three-fourths of the grant [is] to be rent-free for 15 years, after which it shall be assessed at three annas per acre for 10 years, and for 74 years at 6 annas per acre, the whole term being fixed for 99 years."[20] The "liberality," of course, did not end here. At the end of ninety-nine years, the grant was to be "moderately assessed as may seem proper to the Government of the day," and all proprietorial rights of the original grantee, his heirs, executors or assignees were to remain intact.[21] Even grants for "hill forests" were to be made on special terms, though under sanction from the Board of Revenue.[22] To be sure, it was stipulated that, on pain of resumption, one-eighth of the grant be cleared and rendered fit for cultivation at the end of five years, one-fourth in ten years, one-half in twenty years, and three-fourth in thirty years from the date of commencement. In reality, however, these provisions were rarely met as planters and bio-prospectors bought up as much land as possible without regard to their future prospects. Indeed, by 1858–9, even the requirement to show proof of sufficient capital was dropped.[23] As the mania of land-grabbing became the order of the day, vast swathes of Assam's forested and grass-lands came under the speculator's net. The figures were staggering. By 1859, only 7599 acres (or 13.9 percent) of the 54,859 acres in Assam valley and 4000 acres (or 2.7 percent) of the 150,000 acres in Cachar were actually cleared for tea.[24]

Writing a decade and more later, J. W. Edgar, officiating junior secretary to the Government of Bengal (hereafter GOB), harshly criticized the reckless adventurism with which land was taken up during this period. He pointed out that the "inadequacy of demarcating and surveying" in the 1854 Rules "aggravated" the situation beyond control.[25] As we shall see,

---

[19] A. J. Moffatt Mills, Esq., *Report on the Province of Assam* (Calcutta: Gazette Office, 1854), p. 16.
[20] See appendix B, *Papers Regarding the Tea Industry in Bengal*, pp. 96–97.
[21] Ibid., p. 97.   [22] Ibid., p. 97.
[23] Vide, Letter from J. W. Edgar dated September 11, 1873 in *Papers Regarding the Tea Industry in Bengal*, p. xi.
[24] Cited in Siddique, *Evolution of Land Grants and Labour Policy of Government*, p. 21.
[25] See *Papers Regarding the Tea Industry in Bengal*, p. xi.

Conservation or Commerce?

Figure 5.1 View looking over a tea plantation towards a river, Assam, 1860s © The British Library Board, Photo 682(111)

the issue of unmarked (or unclear) boundaries continued to haunt the forest department well into the late-twentieth century. Edgar also made the damning observation that often, "imaginary tracts" of land were sketched up for securing grants that had very little semblance to reality.[26] While the human costs of this uncontrolled expansion – in terms of labor health – have been discussed in Chapter 4, its significance to forest conservancy and silviculture in Assam was lasting and deep.

By the 1860s, however, speculators and planters implored the colonial government to transfer leasehold land grants into permanent holdings on payment of a lump-sum fee. They also protested the stringent provisions of resumption and clearance conditions. Agreed to by Governor-General Canning in October 1861, they were finally revised under orders from the Secretary of State on August 30, 1862. Known as the Fee-Simple Rules, this last modification stipulated that all wastelands on which no proprietorship or known ownership was known to exist were available for sale.

[26] Ibid., p. xi.

Unless otherwise reserved by the Government, any number of lots could be purchased at an upset price of 2 rupees 8 annas per acre.[27] Unhappily for the capitalists, the Fee-Simple Rules also decreed that lands so demarcated be put up for public auction before their final sale. Opposition to this last rule was almost spontaneous. Planters argued that such a move invalidated the time, money, and energy spent on selecting land. The provision of demarcation and survey drawn up by the Fee-Simple Rules was also harshly criticized. As this last requirement fell through, another round of wanton land-grabbing commenced. As Edgar reminisced: "there can be no doubt that the ... insane attempts to extend cultivation, which led to the depression of the tea industry in 1866 and the following years, were very much encouraged by the way in which waste lands were dealt with by the Government."[28] Needless to add, the interests of local agrarian communities, leaseholders and tenants were summarily sacrificed with these freewheeling gestures to the tea lobby.

To be sure, the colonial government went above and beyond these gestures to placate relentless calls to spur the tea industry. Around 1864, the GOB mulled plans to open up lands otherwise meant for ordinary cultivation to tea under a thirty-year grant rule.[29] Designated as areas of "special cultivation," these were earmarked for cash-crops such as tea, coffee and indigo. In the southern Surma Valley, a different set of rules known as the Jangalbari Rules were in place. Under these terms, lands for traditional as well as special crops were doled out with low but progressive revenue rates.[30] By the end of the nineteenth century, even agrarian areas otherwise used by local peasants had come under the planter's purview.

The honeymoon with unchecked land acquisition did not last long. As the tea industry went into a recession around 1866, the GOB tried to recalibrate some of its earlier policies. But the damage had already been done. As lands began to be relinquished and resold, the government suspended the previous withdrawal clause of the Fee-Simple Rules. More to the point, the ability of "native" tea capitalists to open gardens of their own – effectively foreclosed by these wasteland lease terms – was dealt a death blow by this crisis. An Assamese planter, Rosheswar Barua closed down half-a-dozen of his estates after 1868 and the fate of another sixteen "Indian" gardens in the Goalpara district "were [...] nipped in the

---

[27] See *Papers Regarding the Tea Industry in Bengal*, p. xiii.   [28] Ibid., p. xv.
[29] Ibid., p. x and p. xv.
[30] See B. C. Allen, *Assam District Gazetteer*, Vol. 2 (Calcutta, 1905), p. 227; also see Siddique, *Evolution of Land Grants and Labour Policy of Government*, p. 38

bud."³¹ In August 1872, the Government of India (hereafter GOI) directed that sale of lands under the Fee-Simple provisions be suspended altogether. As calls to verify bona fide capitalists became louder, an internal inquiry revealed that "out of a total number of 219 grants, comprising an area of 153,735 acres of land, in 56 only, aggregating 36,288 acres in area, had the necessary clearances been made."³² For Cachar, J. W. Edgar reported to the GOI that a thorough revision would reveal no fewer than 6,00,000 acres taken up for tea planting and left uncultivated.³³ This phenomenon of "locked up" land would have far-reaching consequences for the forest department, as we shall soon see. Edgar also remarked that, owing to large-scale mismanagement, forested areas within these lands – intended to be "reserved" for timber for charcoal and tea-boxes – would be soon found wanting for the purpose. He urged the government to consider "preserving some of its forests with the definite object of supplying [these] at reasonable rates when private resources come to an end."³⁴

By 1874, new political and agro-economic orders were in place. With Assam separated from Bengal with her own chief commissioner, revised wasteland regulations followed. The Leasehold Rules of April 1876 fell back on tenurial rights to land instead of outright sale. Under its provisions, leases were perpetual and gave "a permanent, heritable and transferable right of use and occupancy in the land leased, subject to a payment of land revenue and local taxes and cesses."³⁵ After the expiry of the first thirty years, these lands were to be assessed in accordance with laws of the district in which it was situated. To be sure, the interests of the planting community – bouncing back after the slump of the late 1860s – were never forgotten. Thus, at the time of lease renewal, no portion of these lands were to be assessed at rates higher than that payable for local lands under rice, pulses or other agricultural produce. Though the provision of demarcation and survey were brought back, planters continued to take up lands ordinarily set aside for local cultivators at concessionary rates.³⁶

---

[31] See Amalendu Guha, *Planter Raj to Swaraj: Freedom Struggle and Electoral Politics in Assam, 1826–1947* (New Delhi: ICHR, 1977), p. 17.

[32] See *Report on the Land Revenue Administration of the Lower Provinces for the Year 1870–71* (Calcutta: Government Press, 1872), p. 41

[33] See *Papers Regarding the Tea Industry in Bengal*, p. xvi. [34] Ibid., p. xvii.

[35] Vide *Report on the Administration of the Province of Assam for the Year 1877* (Shillong: Secretariat Press, 1878), p. 32.

[36] This was carried on under the Assam Settlement Rules of 1870. Though planters were required to pay the same revenue rates for ordinary lands as local Assamese *ryots*, they were often released for two-thirds or one-half of regular prices. It was argued that: "tea crop required more time to bear fruit and involved [a] larger outlay of capital than other crops," in *Report on the Administration of the Province of Assam for the Year 1875* (Shillong:

These wasteland settlement rules continued to be in force for the next twelve years. The Assam Land and Revenue Regulation came into effect in 1886 that rearranged land disbursement in three sections. While the lease-holding provisions of 1876 were clubbed under section I, all ordinary lands to be taken up by planters were put under section II (for the Brahmaputra valley) and section III (for the Surma valley). By the end of the nineteenth century, however, the wanton land-grabbing and speculation of the past depleted available wastelands to a bare minimum. The situation was especially acute in the southern Surma valley. Around 1899, even an erstwhile forest reserve in the Longai valley of Karimganj district was thrown open for the industry.[37] As these demands continued to be made of forested areas, run-ins between them and the tea enterprise became more frequent. Wastelands in the northern Brahmaputra valley reached similar states of exhaustion by the end of the Second World War.

Admittedly, the FD was a late claimant in the agrarian landscape of Assam. Even during its first steps under the GOB around 1868–69, officials realized that large tracts of valuable *sâl* (*Shorea robusta*) had already been given over as wastelands to tea companies. And it was not about a single species alone. As the region's first monopoly capitalists, the tea enterprise commanded the early attention of the colonial and provincial administrators. Our foregoing discussion of the wasteland grants amply demonstrate that land – including forest land – was the first, and ongoing, target of this march of capital and commerce. In effect, planters were the biggest landowners of every district in the province. The FD's role and relevance in the province had to be calibrated within this reality. This was to be an uphill task. By 1874, the FD realized that it had to expeditiously *demarcate* its jurisdiction over remaining government forests, and check further *alienation* in the face of wanton land lease and sales.[38] As far as planting interests were concerned, the discourse of "rights" versus "privilege" – mainstay of the department's forest laws – was difficult to sidestep and differentiate. More interestingly, the FD had to confront a peculiar double-bind. If they had to safeguard forests of "value" in the province against misuse, it soon dawned on them that much of this worth was *created* by the tea industry's demands for timber

---

Secretariat Press, 1876), p. 47; also see *Report on the Land Revenue Administration in the Province of Assam for the Year 1880* (Shillong: Secretariat Press, 1881).

[37] See B. C. Allen, *Assam District Gazetteer*, Vol. 2, p. 233; here, "flat" lands were assessed at 15 annas per acre, and "hilly lands" at 3 annas per acres. No revenue was charged for the first three years of possession.

[38] See Gustav Mann's argument in the first *Progress Report of Forest Administration in the Province of Assam for the Year 1874–75* (Shillong: Secretariat Printing Office, 1876), p. 10.

and charcoal in the first place. In a region without woods of universal estimation (teak, for instance) and sparse local usage, I argue that the tea enterprise paradoxically became one of the FD's primary raisons d'être in Assam. It was not a matter of antagonism or approbation alone.

Customer *and* Culprit

> it may be added that forest administration in Assam has become necessary, not so much in order to provide permanently for the requirements of the indigenous agricultural population as to provide the wood and other forest produce required by the extension of the tea industry[39]

Dietrich Brandis, India's first Inspector-General (I-G) of forests toured Assam from February to May 1879. As he moved from one district to the next, the looming significance of the plantation sector for the province – and his department – became increasingly evident. To be sure, Brandis's visit was not the first to make this apparent. Almost a decade earlier, Gustav Mann, under the directions of the Bengal Forest department reported that fine forests along the Dibru river had already been extensively worked by the Assam Tea Company's sawmill.[40] Later, in his first report as the officiating deputy-conservator of forests, Mann noted in 1874 that the Upper Assam Tea Company be held responsible for large-scale felling of valuable trees in the Dibrugarh district. He also remarked that the Balipara sâl forests in the northern Tezpur division had been almost exhausted by tea managers before the forest department could be organized.[41] Of course, the run-ins between officials of the Public Works Department (PWD) and the GOB regarding planter depredations in the Nambor forest during 1859–62 are well explored in the existing literature.[42] Wilhelm Schlich also visited Assam in 1873 and laid out plans for the future management of its forest areas.

Brandis's visit was, however, noteworthy for several reasons. As the first by an I-G to the newly independent province, his recommendations and observations regarding its forest affairs carried significant weight.

---

[39] See D. Brandis, *Suggestions Regarding Forest Administration in the Province of Assam* (Calcutta: Superintendent of Government Printing, 1878), IOR/V/27/560/53, p. 6, British Library, London.
[40] See Arupjyoti Saikia, *Forests and Ecological History of Assam*, especially chapters 1 and 2.
[41] See *Progress Report of Forest Administration in the Province of Assam for the Year 1874–75*, p. 21.
[42] See Saikia, "State, peasants and land reclamation: The predicament of forest conservation in Assam, 1850s–1980s," and Saikia, *Forests and Ecological History of Assam* for a historical assessment of the Nambor reserve experiment in early colonial Assam; also see Richard P. Tucker, "The Depletion of India's Forests under British Imperialism: Planters, Foresters, and Peasants in Assam and Kerala," p. 125.

Moreover, with years of experience in Pegu (Burma), Brandis was uniquely positioned to comment on a region that shared many ecological similarities with its eastern neighbor. This tour also came within a year of the passage of the 1878 Forest Act. With the latter's deep commitment to State authority over forest lands – and the entrenched history of wasteland sales to the tea enterprise in Assam – Brandis faced the full measure of this dilemma during his 1879 tour. The fact that his *Suggestions* spent a considerable portion of its discursive and empirical energies on this question therefore merits attention.[43]

To begin, Brandis commented that aside from sâl, sam (*Artocarpus chaplasha*), and ajhar (*Lagerstræmia reginæ*) from western Goalpara and Kamrup for boat-building, the usefulness of local timber was limited to occasional bamboo works, house-posts, firewood and possible "cabinetry and fancy" furniture.[44] With a poor home market, the economic viability for Assam's timbers seemed to hinge on the ever-increasing demands of the tea industry and its associates (the railways, for instance). For all its sound economic logic, Brandis realized that this was not an unmixed blessing.

The I-G acknowledged that government ownership of rich forest land in Assam was extremely limited. Referring back to the speculative years, Brandis bemoaned the manner in which State rights had been forfeited over vast tracts of land without regard to their necessity or purpose. He countered that out of a total area of 587,409 acres of land held by planters in 1878, only 147,840 acres were put under actual cultivation.[45] As it

---

[43] In addition to the long history of the wasteland leases, the Inner Line Regulation of 1873 added another layer of complexity to forestry administration in Assam. The Inner Line – an ideological, political and spatial separation between the "hill" and "plain" districts of the province was ostensibly aimed at "protecting" the traditional revenue and property rights of the hill "tribes" in eastern Assam (especially beyond the Sibsagar and Lakhimpur districts). However, as historians have argued, these separations were elaborate bureaucratic and legal smoke-screens to define and protect the colonial State's new property order brought in by monopoly capital, especially the tea plantations. The fact that valuable and extensive forests fell beyond the Inner Line made the FD's task of demarcating its jurisdiction in eastern Assam all that more difficult. We do not explore this facet of the province's history in this chapter. For the Inner Line, see Bodhisatwa Kar, *Framing Assam: Plantation Capital, Metropolitan Knowledge and a Regime of Identities, 1790s–1930s*, Unpublished PhD Dissertation (New Delhi: Jawaharlal Nehru University, 2007); Amalendu Guha, *Planter Raj to Swaraj*; H. K. Barpujari, *The Comprehensive History of Assam*, vol. 4 (Gauhati: Assam Publication Board, 1992); Sanjib Baruah, *India Against Itself: Assam and the Politics of Nationality* (New Delhi: Oxford University Press, 1999); and Jayeeta Sharma, *Empire's Garden*, especially Part I.
[44] Brandis, *Suggestions Regarding Forest Administration in the Province of Assam*, p. 4.
[45] Ibid., p. 7; similarly, out of a total area of 1,827,200 acres in the Sibsagar district of upper Assam, planters held 147,071 acres but cultivated only 34,194 acres during this period. Brandis noted that the balance of 112,877 acres was stocked with valuable forests outside the provenance of the FD.

stood, this phenomenon of locked-up lands seriously jeopardized the FD's mandate to husband valuable woodlands for future use. With disdain, the venerable forester remarked: "the result has been that large areas of forest lands have become the property of private persons, and a large proportion of this forest land will probably never be brought under tea."[46] Wastage and laissez-faire aside, the I-G also raised the difficulties of fire protection and freedom of access problems created by these personal estates. As we shall see, issues of jurisdiction and "legitimate" authority – especially in the backdrop of Forest Act VII of 1878 and Assam Forest Regulation VII of 1891 – caused regular run-ins between the tea industry and the FD. To make matters worse, planters paid only one rupee per acre under the Leasehold Rules, and around the same rate for timber standing on it. Thus, the FD received less than 80,000 rupees from lands held in aggregate by the planting community in 1877–78.[47] Recommending "somewhat less liberal conditions" in alienating government hold over its forests than hitherto practiced, Brandis exhorted that all future measures be taken keeping "permanent interests" of both the industry and the country in view.

But his criticisms could only go so far. For even as the FD forfeited proprietary rights over vast swathes of the region's woodlands, it directly benefited – indeed was sustained by – the ever-growing needs of the tea enterprise. For instance, the FD's jump in revenue from 81,568 rupees in 1876–77 to 134,325 rupees in 1877–78 was chiefly due to receipts from planters for trees standing on their lands. Brandis hastened to qualify: "Forest administration in Assam is in no way hostile or antagonistic to the healthy development of the tea industry."[48] But, as the FD's chief spokesman, Brandis proffered a theory of co-dependency that claimed some ground for itself. He observed that in addition to shade and climate, large forest cover provided a natural check on fungus and other insects injurious to the tea plant.[49] Similarly, the wooded areas of western Goalpara district acted as a "screen" against the dry, arid winds of February and March every year and "diminished" its harmful impact on standing tea crops. Beyond ecological rewards, of course, the I-G stressed that keeping the FD close made prudent business sense for the tea industry. He used charcoal and tea-box timbers to make his case. For the former, the I-G presciently observed that till coal-fired machines took over the firing stage of tea production, charcoal would be a *sine qua non* for the industry. He, however, cautioned that its reserves in the province were not endless. Speaking for the Sibsagar district, Brandis argued that the extent of its denudation forced charcoal to be brought in from the

[46] Ibid., p. 9.   [47] Ibid., p. 10.   [48] Ibid., p. 11.   [49] Ibid., p. 9.

neighboring Naga hills. He estimated that for an outturn of 8,282,000 pounds of tea in the district in 1878, close to 100,000 maunds of charcoal were needed. He went on to calculate that this worked out to around 800,000 cubic feet of solid wood. Brandis suggested that under optimum management, a forest area could yield around 50 cubic feet per acre per annum or around 16,000 acres for 800,000 cubic feet of wood.[50] But, as he remarked, the woodlands then supplying charcoal fell far short of this potential. He blamed poor supervision – and overuse in locked up lands – for this outcome. Brandis was also critical of the revenue arrangements regarding charcoal extraction. For him, the rate of 2 annas per maund of charcoal made on government did not equal the value of the wood used.[51] The alternate practice of leasing out portions of government land on pottah for charcoal in lieu of a fixed payment led to greater evils – for planters would often return these only after its stock of wood had been completely extinguished. If charcoal spurred tea manufacture, boxes carried the product to its final consumer. Brandis argued that falling crop prices and a restrictive business climate would soon render the practice of importing tea-box woods from Bengal and Burma unprofitable. Urging planters to consider indigenous woods instead, the I-G reasoned that even this called for demarcating forest areas specifically earmarked for the purpose. I shall return to the tea-box question shortly.

Admittedly, Brandis was up against an intractable paradox. While admonishing the tea industry for all its ills, he had to simultaneously acknowledge the opportunities it created for the FD in Assam. Given its economic footprint and political clout, the I-G also realized that apportioning one-sided blame on the tea enterprise would be both improvident and ineffectual. He thus used a two-pronged approach to make his case. While the logic of co-dependency was one of them, highlighting past errors in economic judgment was another. Brandis thus pointed out that speculative greed – and allowance – in land had bloated the tea enterprise to an "unhealthy" state.[52] In the absence of knowledgeable

---

[50] Ibid., p. 42.
[51] Ibid., p. 46; interestingly, Gustav Mann's proposal to raise charcoal tax from 1 rupee 1.4 annas to 2 rupees in 1874 was deemed to be "rather heavy" by the chief commissioner, see *Progress Report of Forest Administration in the Province of Assam for the Year 1874–75*, p. 24 and Proceedings of the Chief Commissioner of Assam in the Revenue department, October 27, 1875, p. 2.
[52] Ibid., p. 11; commenting on the history of forest conservation in India, Brandis later argued that one of its "special difficulties" in the country stemmed from the activities of "private proprietors." He clarified that by this he mostly meant "European private enterprise, and Europeans in business" who have "as a rule ... one aim only, that is, to get rich as quickly as possible." He hoped that the future development of the Tea and Coffee plantations would bring about a change when those involved in these ventures "would lay greater stress upon maintaining and improving the productiveness, the capital

husbandry, this vast property had been either unused or overworked beyond repair. In calling for leaner estate size, and rational land disbursement in the future, he hoped to pinpoint the policy roots of the conundrum and justify the FD intervention in permanent forest conservancy. After all, Brandis argued, the fiscal benefits of such a move would ultimately benefit the tea industry in the long run. While the I-G's prognosis for parallel growth sounded benign in theory, its implementation on the ground was another matter. As the tea enterprise continued to expand in the last decades of the nineteenth century and beyond, these two sectors were locked in what I call a relationship of productive hindrance.

Your Tree *or* My Tea

> The unlimited reservation by Government of all forest lands is sufficient to throw a damper on any one who is desirous of embarking in tea cultivation.[53]

On their part, planters periodically blamed the colonial government and the FD for creating hurdles in their path. In the aftermath of the 1865 Forest Act, for instance, they found it "unjust" and "unfair" that wasteland sales were denied or curtailed because of its forest cover.[54] Arguing that the tea industry was the primary customer of timber in Assam, they rallied against legal codes that barred them this business privilege. Writing in 1873, F. V. Harvard of the Koliabar Tea Garden remarked that the FD's idea of shielding "valuable" forest resources from the tea enterprise was counterproductive as the latter had created demand for it in the first place. He rallied against the FD's practice of "over-taxing" a "legitimate industry" while allowing *ryots* to fell firewood and strip the country to a "howling wilderness."[55] As the days rolled on, these proprietorial battles between an expanding tea industry and an interventionist FD became frequent and pronounced. The question of jurisdiction and ownership proved to be especially thorny for these two parties. Indeed, as early as 1876, Gustav Mann had pushed for clear demarcation between existing tea grants and adjacent areas under the FD's immediate or intended control.[56] He suggested the use of boundary lines and posts,

---

value of their estates, than upon high dividends," in Sir Dietrich Brandis, *Indian Forestry* (Woking: Oriental University Institute, 1897), p. 36.
[53] James Harris, Sokunbarree Tea Estate quoted in *Papers Regarding the Tea Industry in Bengal*, pp. 64–65.
[54] See *Papers Regarding the Tea Industry in Bengal*, especially pp. 60–93.
[55] Quoted in ibid., pp. 66–67.
[56] See *Progress Report of Forest Administration in the Province of Assam for the Year 1877–78*, p. 3, paragraph 19.

and mounds for the purpose. He also urged divisional forest officers to solicit maps of the tea leases from companies in their districts. The task was easier said than done.

For the FD, securing right-of-way through private estates became a matter of significant concern. This became a pressing issue for forest reserves planned (or already) adjacent to and surrounding existing tea gardens. In such cases, the tacit or active cooperation of the manager was necessary for the movement of the FD's goods, equipment and personnel. Walter J. Mildmay of the Lackwah Tea estate in the Sibsagar district was granted a strip of land in February 1896 through the Sola forest reserve only when he deposed in writing that tract would "only be used for opening up a roadway or tramway and for no other purpose."[57] He had to further guarantee that the provincial authorities would be "entitled at all times to make use of the path ... as far as it passes along the strip of land now applied for when opened."[58] Interestingly, Mildmay's application to disforest 250 acres of this same reserve a couple of years later was rejected by the government.[59] In his defense, the deputy conservator of forests observed that the land in question contained "valuable timber." More crucially, the forester felt that conceding to Mr. Mildmay's request would set a "dangerous precedent" for others and "no reserve would be safe."[60] On June 7, 1899 the Deputy Commissioner of Nowgong (now Nagaon) received a petition from T. Henderson of the Salona Tea Company to disforest around 300 acres within the Diju forest reserve. As it turned out, the portion so demanded had a mixed set-up – low-lying marshland on its south end and mature sâl cover on the hills north.[61] The FD worried that parceling off such a wide area to a private firm would lock-up its access to the sâl, and create tensions regarding right-of-way. After inspecting the land on December 5 with the extra-assistant conservator of forests, the commissioner John Gruning felt that it was in their best interests to let go of the land requested by the Salona Company. Gruning argued that a major section of the low-lying part of those 300 acres were infested with "reeds and jungle" and "worthless" to the FD. For him, opening up that segment for "paddy cultivation" and "coolie

[57] Letter no. 106, Walter J. Mildmay to the Deputy Commissioner, Sibsagar dated Lackwah, February 3, 1896, Assam Commissioner Files-Land Revenue, No. 6 of 1896, ASA.
[58] Ibid., p. 4.
[59] See Assam Commissioner Files – Land Revenue, Collection XIX, No. 30, 1898, ASA
[60] Vide Letter no. B/586, H. Young, Deputy Conservator of Forests, Sibsagar to the Deputy Commissioner, Sibsagar in ibid.
[61] See Assam Commissioner Files – Land Revenue, Collection II, No. 122, 1899, ASA.

settlement" would "benefit the country all round."[62] To allay the FD's fear of loss of control to the sâl beyond, Gruning made Henderson agree to a road up north through his soon-to-be lands that he would build and keep "public."[63]

These transactions were not always smooth. Officers of the Cachar division noted in 1911 that transporting products from the region's unclassed forests posed difficulties due to restrictions on exit through tea estates. Taking cognizance of the problem, the Conservator recommended that even "right-of-way" be reserved while disbursing wasteland leases in the future.[64] Legal stipulations notwithstanding, matters could often spiral out of hand. In one such instance, the manager of Rupacherra tea garden was persecuted and fined 60 rupees by the divisional court in Cachar in 1900. It was reported that he actively obstructed the forest guard and his men from carrying out demarcation work despite the FD's permit through his property.[65] Sometimes, clandestine holdings and legal evasions came to light. An internal inquiry by the revenue department in 1901 "discovered" that the Charaideo tea estate of the Assam Company held close to 238 bighas *as* garden land since 1895 without title.[66] A war of words broke out between Charaideo, the government of Assam, and the FD regarding ownership, land revenue owed, and access to forest resources within this area. The deputy commissioner of Sibsagar, J. C. Arbuthnott reported that the Assam Company consistently "refused to admit" this illegal possession and also barred local *ryots* from using it. For him, the concealed patch – closely guarded by estate "chowkidars" (watchmen) – was "virtually the property of the Company" though not owned by it.[67] Arbuthnott further argued that thatching grass regularly extracted from this undeclared holding was liable for unpaid dues under the Assam Forest Regulation 1891. He urged the government to look into the matter and hold the company responsible. Consider the case of Lieutenant Colonel D. Reid, executive engineer to the PWD of upper Assam, who complained to the government officials in Bengal about the

---

[62] Letter no. 145, Officiating Deputy Commissioner of Nowgong to the Commissioner of the Assam Valley Districts, dated December 7, 1899 in ibid.
[63] Ibid.
[64] See Notes/Group I, Collection No. 4, File 16/1911, Board of Revenue-Survey and Settlement, Eastern Bengal and Assam, ASA.
[65] See *Progress Report of Forest Administration in the Province of Assam for the Year 1901–2*, p. 5.
[66] Vide Assam Commissioner Files-Land Revenue, No. 33, 1901, ASA.
[67] Vide letter no. 2370R dated February 14, 1901 in Assam Commissioner Files-Land Revenue, No. 33, 1901, ASA; though not standardized in usage, a bigha in Assam roughly came to 1338 square meters.

difficulty of acquiring timber from the Nambor reserved forest for departmental use. Among other factors (destruction of forests for opium cultivation, for example), Reid "was convinced that the tea planters, too, were not far behind in damaging the forests, as planters removed trees because too much shade hampered the growth of tea plants."[68]

In some cases, a peculiar mix of irreverence and commercial expediency strained relations between these stakeholders in Assam. Consider the series of exchanges between the Assam Railways and Trading Company (hereafter ARTC) and the Assam government during 1884–8.[69] Set up in 1881 to exploit the mineral resources of the province's eastern sector, namely coal and petroleum, the ARTC bargained out favorable terms for timber felling and railway construction, among other benefits.[70] Under the timber concession (retained for 20 years), the company was allowed to extract, fell, and export wood on a designated tract virtually free for its sawpits, engines, and furnaces, and for erecting necessary buildings. The government retained "beneficial ownership" of the land and no "surface rights" were delegated to the firm. Forest cover, however, proved to be mixed blessing for the ARTC. The area around their Margherita, Tikak, and Ledo colliery was overrun with impregnable jungle and the company desired to clear these areas in the "interest of sanitation" and for providing cultivable lands to its "coolies." In February 1884, ARTC applied for an additional 26,000 acres under the Wasteland Rules of 1854.[71] As pointed out earlier, the 1854 Rules conferred tenure on lease for 99 years, one quarter of the grant remaining revenue-free in perpetuity and rest tax-free for 15 years and then assessed at 3 annas per acre for 10 years, and 6 annas for the remaining 74. The land was thereafter valued at "moderate rates." Under the circumstances, the government was in a quandary. Writing later, the Chief Commissioner, Sir William Ward noted that the ARTC's "extravagant" and "chimerical" demand was not just for business but an ambition to secure "proprietary rights" in land after the expiry of its concessional terms.[72] But given the company's pioneering role in fostering industry in this far-flung region, the government tried to strike a deal. Ward's

---

[68] Quoted in Saikia, "State, peasants and land reclamation: The predicament of forest conservation in Assam, 1850–1980," p. 81.

[69] The following two paragraphs are based on correspondences in File No. 3, Collection V – Immigration, "Return of Inspection of Tea gardens," District Record Room, Office of the Collector and Deputy Commissioner, Dibrugarh, Assam.

[70] See W. R. Gawthrop (compiled), *The Story of the Assam Railways and Trading Company Limited, 1881–1951* (London: Harley Pub. Co. for the Assam Railways and Trading Company, 1951).

[71] See File No. 3, Collection V – Immigration, "Return of Inspection of Tea gardens," p. 2.

[72] Ibid., p. 4.

predecessor, Sir Charles Elliott agreed to the 26,000-acre demand but modified the terms of lease under the revised Wasteland Rules of 1876. Elliott even agreed to "give up the timber valuation" required by the 1876 Rules if the ARTC did not stake compensation in land after the end of its concessions in 20 years.[73] In other words, the Assam government wanted to retain the "rights of State" over lands in the future – even if this meant letting go of revenue in valuable woods.

The ARTC rallied back stating that they were willing to scale down their demand to 20,000 acres, and surrender possession after twenty years. In return, they wanted free access to the timber revenue.[74] Even as the government favored settlement in this requested area, it was unwilling to let go of forests without payment. In fact, when consulted, Gustav Mann valued the timber in this holding at 50 rupees per acre. The ARTC's colonization scheme rested for the next year as Sir William Ward took over the commissionership of Assam.

In January 1886, Captain Beauclerk, Agent to the company sent a note to Ward to "reconsider whether surface rights coterminous with mining rights... could not be granted... at rates which would make it worthwhile to clear the land and seek the establishment of cultivation."[75] Beauclerk changed the tone of his entreaties somewhat. He argued that the renewed demand was to clear the malarial forests around the Company's proposed sawmills from its "unhealthy" influence. He also insinuated that its timber was of no use, and was better sacrificed to its operations. While Ward was willing to concede to the Company's demand for forested land – albeit restricted to 5 square miles in a single compact block – the FD was aghast. Gustav Mann shot back stating that the ARTC had hoodwinked the government on all fronts. He categorically stated that the forests in question was "the most valuable in the whole of Upper Assam" and one of the government's main source of revenue in the future.[76] Mann also recommended against leasing out timber privileges at half-royalty. As even this attempt fell through, the ARTC laid low for the next eight months. In the meantime, the FD proposed to reserve the forested land between the Namdang and Makum rivers – areas squarely within the Company's original demands. Sensing this development, Beauclerk made another attempt to acquire this sector in September 1886. This time, new ideas were floated.

In making this renewed pitch for forest land, the ARTC proposed tea cultivation as one of its main aims. While the Company had not yet added tea to its business profile, Beauclerk made it his primary argument. He also urged the government against reserving the area as it was

---

[73] Ibid., p. 3. [74] Ibid. [75] Ibid., p. 5. [76] Ibid., p. 6.

"eminently" suited for the crop.[77] The Assam government, however, was unwilling and Ward sent back a revised estimate of land that ARTC was allowed to clear. Belligerent and frustrated, Beauclerk wrote to Mann directly in January 1887 suggesting that the Company's Principal Medical Officer had warned of severe "coolie" mortality if the said forests were not immediately felled. Taking affront to Beauclerk's impertinence and "hostility," Ward wrote a lengthy rejoinder to the ARTC. This letter is interesting for several reasons. For one, the Assam government reminded the Company that its original Coal Indenture of 1881 did *not* give it heritable *rights* to destroy contiguous forests even if this was suggested by its medical officer on grounds of hygiene.[78] Ward also argued against alienating "surface rights" as a corollary of – or in conjunction with – the ARTC's business concessions. He made it clear to the Company that these two facets of land revenue were incommensurable and had to be honored accordingly.[79] As Ward left the province in October 1897 and Beauclerk shortly thereafter, the matter came to a long-drawn close.

If the ARTC case was an especially vexed one,[80] the above histories show that sharing personal, productive, and professional space was a complex – even contradictory – process for tea capitalists and the FD in Assam. A. J. Milroy, later conservator of forests in the province, argued that these problems stemmed from the government's myopic land management policies. For him, "past mistakes" of doling out vast tracts to tea created special constraints for the FD.[81] Writing in 1913, he noted both extremes: large plantations without self-sustaining timber source, and vast reserves left unworked and under-used. For Milroy, these mutually incompatible circumstances led planters to look at forest reserves with much disfavor.[82] He urged the government to hereafter apportion smaller holdings for tea, and corresponding areas to conservation depending on need.

Of course, these run-ins and property disputes did not happen in a socio-political vacuum. As mentioned, the availability of wastelands had reached a state of exhaustion by the mid-nineteenth century. As the tea industry continued to grow, the FD bore the brunt of individual and economic pressures to move into areas under the former's control.

---

[77] Ibid. [78] Vide letter no. 1786 dated June 16, 1887, quoted in ibid., p. 13. [79] Ibid.
[80] It has recently been argued that the ARTC was especially problematic as far as the FD's agenda of forest conservation in Assam was concerned; see Saikia, *Forests and Ecological History of Assam*, pp. 131–132. Saikia writes: "the situation worsened to such an extent that the Forest Department had to convince the chief commissioner to threaten the company with discontinuance of contract," in ibid.
[81] A J W Milroy Papers, MSS Eur. D 1054/20, p. 8, Asian and African Studies, British Library, London.
[82] Ibid.

Conservation or Commerce? 153

Sometimes, the department held on to its mandate. Thus, applications to disforest 16,921 acres of the Katakhal Reserve in Cachar were rejected by the FD in 1913, as was a petition to disforest the Lower Jiri Reserve.[83] But by 1919–20, the pressures of the industry – and the needs of agrarian and peasant settlement – forced the FD to open its unclassed state forests to all leases, including tea.[84]

Trees *for* Tea: The Politics of Codependency

No tea estate can afford to be without forest.[85]

Departmental wrangling around jurisdiction and property rights notwithstanding, deeper economic logics connected the FD and the tea industry. Let us look into this a little closely.

Around 1883, an important case was placed before English jurist and politician Sir Robert Phillimore in London. The plaintiffs – merchants from Mincing Lane – brought the owners of *Asia* to court arguing that the "foul state of the ship" or the "nature of cargo carried" had "tainted" and "impregnated" the loose tea with a flavor and rendered them unsalable.[86] Upon several days of hearing, and on the opinion of experts, the judge declared that the wood of the chests in which tea was transported was responsible for the damage. It was chemically demonstrated that carbonate of lead had formed on the lining of the chests – and that this was induced by "acidity in the wood enclosing the Tea."[87] Further analyses suggested that the damage was "greatly aggravated by the newness of the wood, which was in far too green a condition, and gave forth a strong aromatic smell."[88] For the tea enterprise, the importance of wood extended far beyond charcoal fire.

Though the Keating Saw Mill, and the Dehing Saw Mill had already begun supplying tea-boxes to the plantations in Assam from 1874 to 1875,[89]

---

[83] See *Progress Report of Forest Administration in the Province of Assam for the Year 1913–14*, p. 1.
[84] See *Progress Report of Forest Administration 1919–20*, p. 2; for an analyses of the political and social ramification of these run-ins – especially in terms of peasant revolts and land settlement in the 1980s – see Arupjyoti Saikia, "State, Peasants and Land Reclamation: The Predicament of Forest Conservation in Assam, 1850s–1980s."
[85] Claud Bald, *Indian Tea: Its Culture and Manufacture, Being a Textbook on the Culture and Manufacture of Tea*, second edition (Calcutta: Thacker, Spink and Co., 1908), Chapter XII, p. 158.
[86] See Section XX, Part I, "Injury to Tea From the Use of Improper Woods for Tea Chests," reprinted in *The Tea Planter's Vade Mecum* (Calcutta: Office of the Tea Gazette, 1885), p. 213.
[87] Ibid., p. 213.   [88] Ibid., p. 213.
[89] Vide *Progress Report of Forest Administration in the Province of Assam for the Year 1875–76*, p. 23.

planters were wary of "poisonous woods" or an experience similar to *Asia* affecting their supplies.[90] Throughout the late nineteenth century and into the twentieth century, the question of wood variety – and thereby the use of local forests – became a point of debate between the planting community and the FD. The native Simul tree (*Bombax malabaricum*) emerged as a clear choice, and its wood continued to be favored by planters during this period. Close to seven mills operated during 1910–11 in the Lakhimpur district alone with an outturn of 546,919 boxes.[91] The insistence of numbers alone, however, misses the complexity of the problem. Given the disparate geographical location of Assam's forests and saw mills – some close to the Brahmaputra river and others far inland – a comparatively strong wood was needed that could withstand the rough and tumble of its journey to Calcutta and beyond. This, too, was a paradox in itself as freight charges were directly proportional to weight. In other words, a species that met all these requirements, hardy, durable, and light worked best for tea-boxes but was difficult to come by. Though Simul fitted the demands perfectly, the operation had its hazards. Usually felled during the cold weather, its logs were transported to mills to be cut into shooks – actual pieces from which the boxes were made. These shooks were then sewn up and stacked on end out in the open to dry. But time was of the essence, as an extended drying period caused them to rot and split. The depredations of the small beetle, locally known as "Ghong" and "probably a species of the Bostrichus" added to the headache.[92] A. Smythies, writing for the *Indian Forester*, noted that outside of the busy months of June–September, a "large stock [was] liable to accumulate, and [that] there [was] always a good deal of loss."[93] It was only by seasoning the shooks that the dangers of rotting could be offset. During seasons of heavy demand, unseasoned boards were often sent to gardens, and its managers expected to complete the drying process. Failure to do so left a "cheesy" odor in the planks, which then extended to the packaged tea. Transporting these shooks was also a double-bind; in a province with skeletal communication networks, the Brahmaputra was the fastest route to do so during much of this period. However, as Smythies noted, submerging shooks in water for an extended period canceled out the seasoning and led to discoloration. In many instances, woods

---

[90] See "Injury to Tea From the Use of Improper Woods for Tea Chests," reprinted in *The Tea Planter's Vade Mecum*, p. 213–214.

[91] Vide F. Beadon Bryant, *A Note of Inspection on Some of the Forests of Assam* (Simla: Government Monotype Press, 1912), p. 10, Asia & Africa P/W 453, British Library, London.

[92] A. Smythies, "Note on the Use of Simul Wood for Tea Boxes in Assam," *The Indian Forester*, Vol. xx, No. 10, October 1894, p. 363.

[93] Ibid., p. 364.

Conservation or Commerce? 155

from other trees – notably Kadam (*Anthocephalus kadamba*) or Makai (*Shorea pennicellata*) – were mistakenly brought in for Simul.[94] The former was reported to have a "most unpleasant odor like rancid butter" and most unsuitable for tea-box production. As far as Kadam was concerned, these "expert" opinions often clashed with on-the-ground observation. S. E. Peal, commissioned to write a series of articles on the tea-box question for the *Indian Tea Gazette* in 1883, provided contrary suggestions on Kadam's suitability for making shooks. For Peal, the tree was "one of the best" for tea-boxes and had a remarkably quick growth for its first six to eight years.[95] He encouraged fellow planters and the government to use the Kadam, both for foliage and as wood for tea-boxes, given its robust natural growth in Assam. F. Beadon Bryant, I-G to the FD, wrote in 1912 that planters did make use of *Anthocephalus kadamba* though to a far lesser extent than Simul.[96] Interestingly, Peal also recommended the sam (*Artocarpus chaplasha*) for tea-boxes though he recognized its local usage for muga silkworms and consequent restrictions on felling and availability.[97]

If Simul remained the planters' wood of choice, contributing to more than "80 percent" of tea-box production in 1911–12, the conundrum of local usage was far from resolved.[98] For Simul had other claimants besides the planting community. As Beadon Bryant reported, a large part of the vexation came from *Bombax malabaricum*'s value as fuel and firewood to the "native" Assamese. In addition, shooks made from Simul could be sold as "dry goods packing cases" in Calcutta, London, and other "major centres of consumption."[99] Amidst these competing demands, Beadon Bryant argued that despite "satisfactory" progress, the local tea-box industry in the province was in "a shaky state."[100] Bemoaning its failure to fully meet the tea industry's demand, the I-G noted that the imported Venesta box made of Norwegian, Swedish, and Russian wood continued to be imported.[101] He underscored that the Venesta's durability and immunity from insect attacks,

---

[94] Ibid.
[95] S. E. Peal, "Indian Timbers for Tea Boxes and Other Purposes," for the *Indian Tea Gazette*, reprinted in *The Tea Planter's Vade Mecum*, pp. 222–241.
[96] See F. Beadon Bryant, *A Note of Inspection on Some of the Forests of Assam*, p. 10; Bryant reported that planters also made use of Dumbail (*Bombax insigne*), Tulla (*Tetrameles nudiflora*), the Ram Dala (*Duabanga sonneratioides*), Hollock (*Terminalia myriocarpa*) and Aam (*Mangifera indica*) among other species for making tea-boxes, though Simul overrode all of these in terms of preference.
[97] S. E. Peal, "Indian Timbers for Tea Boxes and Other Purposes," p. 239.
[98] See F. Beadon Bryant, *A Note of Inspection*, p. 10.   [99] Ibid., pp. 11–12.   [100] Ibid.
[101] Beadon Bryant pegged the number of imported tea-boxes to Assam (by sea) at 172,361 for 1906–7; 197,208 for 1907–8; 260,350 for 1908–9; 332,820 for 1909–10 and 283,310 for 1910–11, in ibid., p. 10.

and ease of storage made the imported product vastly more desirable than its indigenous competitor. Of course, the fact that these "foreign" shooks returned a higher commission for tea agents than their domestic counterparts was well known, though planters rarely acknowledged it in public.[102] The paradox played out once again. As late as 1912, Beadon Bryant realized that outside the tea industry, the marketability of Assam's forest produce was limited and negligible. He remarked: "in traveling through the forests ... the point that struck me most forcibly was [its] vastness ... and the small quantity of material [...] brought out of them."[103] For the I-G, lack of knowledge about the province's forest resources, geographical inaccessibility, the Inner Line restrictions, and unestablished markets for its timbers led to forestry's "undevelopment" in the region. Though "reserved forests" covered only about 8.7 percent of Assam's total land area in 1910–11, the I-G saw no reason to extend it given these aforementioned conditions. In such a scenario, the tea enterprise – and its subsidiary needs including tea-boxes – seemed to provide legitimacy to the FD's continuing investment in silviculture and conservancy. But it was not just the FD that benefitted from this arrangement. The tides of fortune could often turn in the other direction. In one such instance, H. C. Hill, officiating I-G, noted in 1896 that in the face of dwindling tea sales, certain gardens in the Kamrup district were almost wholly subsisting on the sale of *sâl* trees within their estates.[104]

To be sure, much had already been conceded as far as spurring the local tea-box industry (and charcoal) was concerned. The category of "unclassed state forests" – a regional innovation of the Assam Forest Regulation of 1891 – was created partially to cater to this demand. Indeed, asking for a provincial law separate from the central act captured the peculiar, and somewhat ironic, odds of the FD in Assam. Writing in December 1889, F. C. Daukes, officiating secretary to the Chief Commissioner, entreated the Revenue Secretary, GOI that the 1878 act was "not in conformity with local conditions."[105] He argued that outside of government grants, vast tracts of land were still available in the province whose use could not be regulated only within the "reserved" or "protected" categories of the 1878 act. Finding these constraints "cumbrous," Daukes called for greater powers to the local government to legislate

---

[102] Ibid., p. 12.   [103] Ibid., p. 8.
[104] See H. C. Hill, *Note on an Inspection of Certain Forests in Assam* (Calcutta: Office of the Superintendent of Government Printing, 1896), p. 8, IOR/V/27/560/55, British Library, London.
[105] Vide letter no. 4242/13, dated Shillong, December 7, 1889 in IOR/L/PJ/6/313.

timber and forest produce movement, or their use, beyond these two areas.[106] In fact, I-G Berthold Ribbentrop had expressed similar concerns during his visit to Assam that year.[107] What was left unstated in these exchanges was the FD's eco-legal dilemma in the province. For not only did Assam not have endless valuable woods and established markets for its timber as Brandis indicated earlier, much of this wealth was either in the hands of – or generated by – *zamindars*, private owners, and tea planters. Statistically speaking, the FD only owned a very small portion of this as "reserves" as already mentioned. In asking for greater local powers, therefore, the Assam authorities were seeking autonomy to circumvent these conundrums in a more effective and pragmatic way. Viceroy Lansdowne gave his assent to this request, and the Assam Forest Regulation came into effect from December 1891. Modeled on the Upper Burma Regulation of 1887 rather than its 1878 Indian counterpart, the Assam act ensured that sufficient leeway was left for its local economic exigencies to operate. Under section 28 (1) of the Regulation, for instance, the local government (with the previous sanction of the Governor-General in Council) could declare previously reserved areas open.[108]

If expedient deforestation was an undeclared aim of the 1891 act, the category of "unclassed state forests" gave this ambition a name. While addressing other pressing issues such as peasant settlement and revenue generation from grazing, *jhuming*, and rubber extraction, these unclassed forests also aimed to settle the FD's dilemma vis-à-vis the tea industry. In other words, these tracts were catchments of forest wealth for the tea enterprise (and others) to dip into without disturbing the FD's authority and control over permanently "reserved" areas. Thus, in 1903, out of a total area of 22,287 square miles under the FD's control, 18,509 square miles were unclassed forests whereas only 3778 square miles were "reserved."[109] The actual working of this category, however, revealed further ironies and contradictions. For one, a large

---

[106] Ibid., p. 2.
[107] Ribbentrop remarked: "whilst they are constituted the greater bulk of the forests remain unprotected, in so far that the local government can make no provision protecting certain classes of trees, or for effectively regulating the extraction of forest produce," *Note on An Inspection of the Forests of Assam* (Simla: Government Press, 1889), paragraph 67, quoted in Saikia, *Forests and Ecological History of Assam*, p. 125.
[108] See IOR/L/PJ/6/313, p. 10; also see *The Assam Forest Manual*, Vol. I (Shillong: Government Press, 1923), especially Part II, pp. 41–86. The 1891 Regulation was later modified in light of the new India Forest Act XVI of 1927, and amended again in 1931 and 1933.
[109] Quoted in W. Schlich, *Manual of Forestry*, Vol. I (London: Bradbury, Agnew & Co., 1906), p. 111.

portion of the unclassed state forests came from lands relinquished by tea estates in lieu of privileges inside reserved woodlands. By one estimate, close to 205,066 acres of unclassed forests in 1897 came from such exchanges.[110] The practice continued well into the second decades of the twentieth century.[111] However, in most cases, long possession of these now freed-up lands by plantation companies meant that they had been overworked or completely depleted of its forest capital. As A. John Harrison, manager of the Meckla Nadi Saw Mills in the Lakhimpur district, admitted to the deputy commissioner of forests in March 1912: "Semul, which formerly existed in great abundance along all the river banks, in close proximity to the mills, had been exploited or exterminated in many cases."[112] In other cases, the cultivating habits of ex-tea garden laborers were blamed for driving this class of forests to exhaustion. J. C. Arbuthnott, the Commissioner of the Surma Valley districts reported to the Secretary of the Board of Revenue in 1911 that former "coolies" regularly worked the unclassed forest areas for growing sugarcane and other crops and abandoned them when used up.[113] As the conundrum came full circle, the FD looked to establishing new "reserves" or tapping into already existing reserves to husband Simul for tea-boxes. Indeed, Bryant strongly and urgently recommended such a move in the face of "dwindling" resources in the unclassed forests.[114] In praxis, the *idea* of earmarking "reserves" in the interest of private capitalists was riddled with inconsistencies. For, as B. H. Baden-Powell argues in *Forest Law*, the first class of State or Reserve forests were "tracts par excellence," wholly possessed by the government, whose "boundaries" had been "fully and authoritatively demarcated," and where "private" rights had been "settled, extinguished or prohibited" from "prescriptive" future growth.[115] But the 1891 Regulation was specifically aimed at resolving – or at least mediating – these challenges in Assam. The ability to turn to unclassed state forests, to create new reserves, for Simul illustrates a case in point. In a region of alienated and locked-up private

---

[110] Vide *Progress Report of Forest Administration in the Province of Assam for the Year 1897–98*, p. 1, quoted in Saikia, *Forests and Ecological History of Assam*, p. 95.
[111] See *Progress Report of Forest Administration in the Province of Assam for the Year 1913–14*, p. 1.
[112] Quoted in F. Beadon Bryant, *A Note of Inspection*, appendix, p. 15.
[113] Letter No. 4148, Collection No. 4, File 16/1911, Board of Revenue-Survey and Settlement, Eastern Bengal and Assam, ASA.
[114] See F. Beadon Bryant, *A Note of Inspection*, p. 2.
[115] B. H. Baden-Powell, *Forest Law: A Course of Lectures on the Principles of Civil and Criminal Law and on the Law of the Forest; Chiefly Based on the Laws in Force in British India* (London: Bradbury, Agnew & Co., 1893), Lecture XVI, especially pp. 225–278.

lands, and limited local demands, such legal maneuvers seemed mutually necessary.

But the relationship of co-dependency did not always produce desirable financial results. Indeed, the outcome could be decidedly contradictory. For instance, a total of 946,095 logs felled in the Cachar Division in 1902–3 fetched the FD 26,870 rupees whereas a corresponding felling of 691,094 in 1901–2 brought in 27,968 rupees for the department.[116] Upon inquiry, E. S. Carr, the Conservator reported that the reduction in 1902–3 – despite a higher outturn – was due to the large number of "third-class timber" exploited for the tea-box industry.[117] C. E. Muriel, the Conservator of Forests, reported during 1904–5 that "unregulated arrangements" with saw mills for procuring Simul had led to wanton "wastage" and "rotting" in the forests of Lakhimpur.[118] More to the point, the FD's desire to offset "foreign" competition by remitting tea-box royalty starting August 22, 1912 caused it to lose 30,000 rupees that very same year.[119] As the impending war, and revenue concerns kept the remittance game on and off,[120] the FD admitted that the present royalty system was "faulty" and led to "wasteful conversion."[121] From 1919 to 1920, a "log-based" rather than a box-based royalty method was introduced. For all their claims, the FD and the tea industry could rarely have it both ways.

"Truth" and Reality

> whatever areas are capable of *yielding value*, are sure to be appropriated, and more or less jealously guarded, by *some* one – either by individuals or by the State itself.[122]

The FD's mandate in British India was predicated upon some "forestal truths" vis-à-vis competing stakeholders.[123] Effected through notions of

---

[116] Vide *Progress Report of Forest Administration in the Province of Assam for the Year 1902–3*, p. 16.
[117] Ibid.
[118] See *Progress Report of Forest Administration in the Province of Assam for the Year 1904–5*, p. 8.
[119] Vide *Progress Report of Forest Administration in the Province of Assam for the Year 1912–13*, p. 15; also see Proceedings of the Chief Commissioner of Assam in the Revenue department, No. 201 R dated January 12, 1914 in ibid. p. 2.
[120] The tea-box royalty system was withdrawn on September 1, 1914 and later reinstituted due to pressures from the tea industry. It was finally withdrawn from January 1, 1915 though a 20 percent ad valorem reduction in royalty rates was again sanctioned in 1927.
[121] See *Progress Report of Forest Administration in the Province of Assam for the Year 1918–19*, p. 13.
[122] B. H. Baden-Powell, *Forest Law*, p. 2; italics in the original.
[123] The phrase "forestal truth" was used by Baden-Powell in "Forest Conservancy in its Popular Aspect," *Indian Forester*, Vol. II, No. 1, July 1876, p. 8.

ownership, property, and law, these "truths" – or principles – guided its dealings with indigenous forest dwellers, peasants, *zamindars*, other agrarian communities, private capitalists and other arms of the colonial State. Though this well-documented history[124] need not be repeated, examining some of these "truths" helps us better understand the genealogy of the problem discussed above.

It is interesting to note, for instance, that the FD's institutional origins in the subcontinent began with efforts to remedy the devastation caused by private contractors. Indeed, its establishment in 1864 was made urgent by the rapid depletion of valuable teak (*Tectona grandis*), sâl and deodar (*Cedrus deodara*) for railway expansion in peninsular and western India. If laissez-faire was the order of the day, the idea of limiting use (and overuse) of forest resources was novel and anachronistic. Gregory Barton suggests: "the realization of market value ... meant the forest itself had to be proclaimed, demarcated, regulated, and policed ... to see the market value of the forest meant the imposition of a 'forest conscience' on the minds of the local inhabitants, as well as the merchants and the government of India."[125] To be sure, this was a theoretical paradox and a substantive challenge. If forest "worth" could only be guaranteed, generated and sustained by restricting use, it entailed changing the *nature* of the relationship between patron and client. And this meant not just European prospectors, but agrarian communities, peasant groups, landlords and village dwellers who worked India's woods for centuries. In this crowded group, the FD needed a two-pronged maneuver to stake its claim as the forests' "true" caretaker, owner, and access regulator. It has to extinguish or redefine previous modes of forest use and simultaneously invest them with new rules and limits. But primarily, the FD had to demonstrate that forests – or some forests – were its exclusive domain and outside the reach of these other claimants. As Bengal civil servant and one of the architects of India's forestry system, Baden-Powell argued: "Forest conservancy starts from a basis of property. You cannot conserve

---

[124] See Ramachandra Guha, "Forestry in British and Post-British India: A Historical Analysis," *Indian Economic and Social History Review*, Parts I & II, Vol. 18, No. 44 (October 29, 1983): 1882–1896 and *Indian Economic and Social History Review*, Parts III & IV, Vol. 18, No. 45/46 (Nov. 5–12, 1983): 1940–1947; Ramachandra Guha, *The Unquiet Woods: Ecological Change and Peasant Resistance in the Himalaya*, and Guha, "An Early Environmental Debate: The Making of the 1878 Forest Act," *Indian Economic and Social History Review*, Vol. 27, No. 1 (1990): 65–84; also see Mahesh Rangarajan, *Fencing the Forest*, S. Ravi Rajan, *Modernizing Nature*, K. Sivaramakrishnan, *Modern Forests*, Richard P. Tucker, *A Forest History of India* (New Delhi: Sage Publications, 2012), and Gregory A. Barton, *Empire Forestry and the Origins of Environmentalism* (Cambridge: Cambridge University Press, 2007).

[125] See Gregory A. Barton, *Empire Forestry and the Origins of Environmentalism*, p. 75.

Conservation or Commerce? 161

a Forest or the Forest area of any district, unless you have either an absolute or a more or less limited proprietary right in it."[126] For this vision of "Pax Sylvana,"[127] a regime of what I call eco-legality was pressed into service. Though "conservancy" had already existed in some form or the other since the late 1830s, the first "official" legislation came with Act VII of 1865. As the first attempt to endow the State with juridical rights over woodlands, the 1865 act was feeble in its conception and tentative in matters of enforcement. While it called for earmarking "government forests" over "lands covered with trees, brushwood or jungle," it also stipulated that such laws would not "abridge or affect an existing rights of individuals or communities."[128] As time went on, a more robust and decisive mechanism to protect woodlands was felt. Baden-Powell was categorical that the 1865 act had dealt with "rights of [individual] users" in a far too liberal and cavalier fashion.[129] At a Forest Conference in 1874, he proposed an "annexationist"[130] plan for government control over India's forests. Baden-Powell's call turned into Act VII of 1878, a wide-ranging legislation that stamped the State's authority over the country's forested areas. But to do so, it used what has been termed as a "legal sleight of hand."[131] Ramachandra Guha argues:

> [the 1878 Act] sought to establish that the customary use of forest by the villagers was based not on 'rights' but on 'privileges' and that this 'privilege' was exercised only at the mercy of the local rulers. Since the British were now the rulers, the rights of absolute ownership were held to be vested in them. As one officer bluntly stated in 1873, "the right of conquest is the strongest of all rights – it is a right against – which there is no appeal". The 1878 Act was the means by which the success of this [...] was assured.[132]

If colonial annexation was the *locus standi* for the State's authority over forested areas – and the removal of "traditional" claims to the same – its laws needed to specify the mechanisms of control. The 1878 Act thus engineered a tripartite division of wooded tracts into "Reserved," "Protected," and "Village" forests.[133] Prized assets, the first category

---

[126] Baden-Powell, "The Political Value of Forest Conservancy," p. 280.
[127] This is Rangarajan's term; see his *Fencing the Forest*, p. 5.
[128] See *The Unrepealed General Acts of the Governor General in Council*, Vol. II [1864–1871] (Calcutta: Office of the Superintendent of Government Printing, 1876), pp. 688–689.
[129] B. H. Baden-Powell, *Forest Law*, p. 188.
[130] This is Guha's term; see his "An Early Environmental Debate: The Making of the 1878 Forest Act," p. 67.
[131] Guha, "Forestry in British and Post-British India: A Historical Analysis," Parts I & II, p. 1884.
[132] Ibid.
[133] See *The Unrepealed General Acts of the Governor General in Council*, Vol. III [1877–1881] (Calcutta: Office of the Superintendent of Government Printing, 1898), pp. 124–156.

was intended to include trees of the highest fiscal estimation – an area "par excellence," as Baden-Powell put it.[134] The FD attempted to permanently settle and control reserved areas by either removing existing rights, transferring them to "protected" forests, or by retaining minimal rights in exchange for a fee. The second class of woodlands was a buffer zone for the FD. Though controlled by the State, rights in them were recorded but not settled. In other words, claims could spring up in "protected" forests but not in "reserved" tracts. As the government clarified: "as regards the people, the chief difference is that ... in a reserved forest everything is an offence that is not permitted, while in a protected forest nothing is an offence that is not prohibited."[135] There were some restrictions even in "protected" woods, however. For instance, certain tree species in these segments could be reserved if necessary, and portions closed to grazing and firewood collection. The third category was technically the preserve of local village communities, outside State supervision, though these were in very limited operation throughout the period under review. The 1878 Act also stipulated detailed instructions for timber extraction, valuation, felling, and transportation. In addition, fines for trespassing, fires and other "illegalities" were also drawn up. Lastly, the provincial governments were allowed to institute subsidiary laws in conjunction with this central act as and when deemed fit. Except a few local governments – Burma, Madras, and Assam, for instance – most provinces fell in line with the 1878 Act. It had a long career and was only replaced by the Indian Forest Act XVI in 1927.

These sylvan visions – and their means and ends – reveal some interesting logics as far as our story is concerned. Though difficult to generalize on an all-India basis, the primary aim of this new regime of "scientific" forestry was to rationalize land use and maximize revenue. As Mahesh Rangarajan argues: "tensions were centered on issues of access to forests and control of land and trees rather than on maintaining a particular mix of species."[136] If recalibrated land management formed the basis of this new imperial department, its defining principles made several assumptions. For one, the 1878 Act left "forests" – as category and concept – undefined. Sensing the potential misuse of the term if *fixed* in perpetuity, Baden-Powell argued that, for the purposes of law: "it was sufficient to provide steps for demarcating certain areas and constituting them Forest Estates ... [and then] subject them to law."[137] If a still "wider meaning was required," he recommended that the term "forest or waste land,

---

[134] B. H. Baden-Powell, *Forest Law*, p. 229.
[135] See "Resolution on the Forest Policy of the Government of India," reprinted in *The Indian Forester*, No. 11, November 1894, pp. 414–422.
[136] Rangarajan, *Fencing the Forest*, p. 202.   [137] Baden-Powell, *Forest Law*, p. 199.

[i.e.,] everything that is not actually cultivated field or meadow land" be used. This looseness had several implications. Legally speaking, it meant that State ownership of forest land was not a priori sanctioned but came into existence *after* such lands were identified, bound, and appropriated. Baden-Powell's legalese thus hinted that the FD's power – indeed legitimacy – stemmed not *from* law, but from its ability to secure designated areas which then came under its ambit, a posteriori.[138] In other words, the department started with the idea that there were enormous tracts of woodlands – unburdened with legal ownership rights *by* the colonial State – that it could take over, at will or by force, and render them "valuable." If the long history of peasant protests against the FD in India shows the coercive side of this ambition, the department's run-ins with planters in Assam highlight that not all landscapes were theirs for the taking, de facto. Indeed, this chapter shows that the FD's jurisdiction, capabilities and authority were severely strained in Assam due to other colonial claimants.[139] As far as these logics of property were concerned, wastelands created special problems for the FD. Despite Baden-Powell's claim that "the right of the State to all waste lands not lawfully occupied, nor forming part of recognized estates, is, for Forest officers, the most important head,"[140] the fact remained that much of it had been squandered under Lord Canning's liberal rules for special cultivation such as tea, coffee, and cinchona. Revising his *Lectures on Forest Law* in 1893, Baden-Powell agreed that, despite later revisions to lease rather than sell wastelands, Canning's "questionable policy" had "sacrificed" much of the State's surface rights over these areas.[141] Our foregoing discussions fully bear out the implications of this history. To be sure, it is ironic that Governor-General Dalhousie would agree to the Old Assam Rules of 1854 just a year before his famous Minute to prevent "all unthrifty management of the forests, on which we must mainly depend for the supply of necessary timber."[142]

---

[138] Rangarajan argues: "the reserved forest was thus a juridical category, not a descriptive one. The creation of extensive government forests had been 'not so much for purposes of forestry,' as for the alienation of property rights to land. Many areas were annexed because there was 'no one' whom the government wished to recognize as proprietor," in *Fencing the Forest*, p. 74.

[139] To be sure, private forests – especially in the *zamindari* and *malguzari* tracts – did not come under the purview of the forest acts. Though the government later tried to burden them with timber duties (among others), these were beset with problems and led to frequent clashes; see Rangarajan, *Fencing the Forest*, chapter 2 for an assessment.

[140] Baden-Powell, *Forest Law*, p. 211.   [141] Ibid., p. 216.

[142] See "Copy of a Minute by the Marquis of *Dalhousie*, Dated the 28th Day of February 1856, Reviewing his Administration in India, from January 1848 to March 1856," p. 23 in House of Commons Parliamentary Papers, May 30, 1856.

If the FD was keen to impart new forest lessons in resource management and land use, these parables had to be adjusted keeping local contingencies in mind. Even law could be bent or circumvented if needed. Our discussions of the Assam Forest Regulation VII of 1891 show that deviating from its ideological principles was imperative – even expedient – for the FD's legitimacy and existence in this region of empire. Indeed, in this shared field of power and property, excessive regulations could have decidedly contrary outcome. W. Cooper, Secretary to the Surma valley branch of the Indian Tea Association told the President of the Assam Forest Enquiry Committee in 1929 that villagers around his company's gardens preferred to buy items of daily use from him because of the FD's excessive rules regarding their removal from government land.[143] It is ironic that Baden-Powell used the analogy of zealously guarded private estates in making his case about strong State guardianship of forests. The venerable civil servant rhetorically asked: "in what sort of respect would that man be held, who allowed his fences to be broken down, his fruit to be gathered, and his fields to be trampled, and that in defiance of notice boards prohibiting in stately terms the trespass, and assuring everybody that he was the sole owner?"[144]

As far as the politics of co-dependency between the FD and the tea industry in Assam was concerned, these questions could not have straightforward, one-sided answers for the most part, as we have seen.

---

[143] W. E. D. Cooper to J. Hezlett, dated Binnakandi, January 19, 1929 in *Evidence Recorded by the Forest Enquiry Committee, Assam for the Year 1929* (Shillong: Assam Government Press, 1929), p. 2, ASA.
[144] Baden-Powell, "The Political Value of Forest Conservancy," p. 283.

6   Plant and Politics

> The main danger the authorities have had to guard against is that the ideas of the non-cooperation party might widely *infect* the tea garden coolies. Any such general *infection* would be a most serious menace to the tea industry [and] render the present methods of personal management impossible.[1]

The years 1920–21 were turbulent years for the Assam tea enterprise. While the winds of Gandhi's noncooperation movement blew across India, a wave of worker protests and walkouts – starting in the Assam valley and then moving southward to Cachar and Sylhet in the Surma valley – crippled plantation operations during these years.

The political timing was fortuitous, for much had already been rotten in the state of Assam tea in the run-up to these events. A wartime bubble, and government price assurance, had propped production beyond sustainable limits. As stocks piled up and profits plunged after 1919, the Indian Tea Association (hereafter ITA) and planters devised an invidious strategy to manipulate law, crop agronomics, and wage structure to tide over the crises. To be sure, these extra-market and extra-legal policies – the culture of commerce, as I call them – had been in existence for decades; the postwar scenario only exacerbated its implementation. Faced with retrenchments, limited work, and subsistence earnings as a result of these plantation tactics, laborers finally erupted in anger and frustration in the autumn of 1920–21.

These protests were remarkable *not* because of their timing or Gandhian oeuvre. Indeed, nationalism, Congress "infection" or labor unionism are unhelpful historical epithets to categorize these events. Instead, this chapter shows that the significance of these walkouts was in that it forced open the illegal and unseen market logics that ran the Assam plantation industry. In this regard, its impact and significance was

---

[1] *The Daily Telegraph*, March 24, 1922, emphasis added. See IOR/L/PJ/6/1798, file no. 1859, Asian and African Studies, British Library, London.

Figure 6.1 Group portrait of field laborers on a tea plantation, Assam, 1860s © The British Library Board, Photo 682(110)

beyond doubt. As the edifice of indenture crumbled in the post-1920/21 period, these strikes – and the recorded and unpublished government reaction to the crises – demonstrate that an unregulated and manipulative wage and work structure had become the order of the day in these estates. The theory and praxis of that regime, its antecedents, and *why* they converged in the making of systemic labor impoverishment (and the 1920–21 walkouts) is the focus of this chapter. I also suggest that the hermeneutics of causality – whether colonial, nationalist, or contemporary – is a limiting historical tool to analyze these walkouts. It revealed much more than a tussle between means and ends.

"History Reads Workers"

On September 6, 1920, workers of the Hansara tea estate of the Doom Dooma Company in upper Assam went on strike. An official report of the time indicated that this was not a garden-wide phenomenon, with "coolies" demanding better quality (and adequate supplies) of rice.[2] About a week later, in the neighboring Raidang estate, the situation turned more serious. Here, workers allegedly beat the garden *jemadar* to death, and also "pelted" two junior Assistants. Around September 21, on the Pabhajan tea estate, a "mob" surrounded the manager's bungalow and "flung flower-pots into the verandah."[3] Next day, a similar case was noted on Dhoedaam, an adjoining garden under the same management. The Superintendent of Police, Mr. Furze, dispatched along with two constables to quell an increase in "disturbances" were purportedly assaulted and "severely injured" by angry workers. The officiating commissioner of the Assam Valley districts noted that the main worker demands at Dhoedaam was to increase the supply of rice (then sold at 3 rupees a maund) to 8 seers a week from 6 and their pay from 6 to 8 rupees per month.[4] The Deputy Commissioner (hereafter DC) touring the estates impressed upon the manager to "raise the allowance of rice at once to 7 seers a week."[5] He further announced that the question of rise of pay was to be given a hearing at the general meeting.

Promises notwithstanding, around forty workers were arrested at Pabhajan and Dhoedaam estates on September 25, 1920. In retaliation, around four hundred workers left Dhoedaam and marched into the town of Dibrugarh and were eventually persuaded to return back by the DC. Meanwhile, at the Samdang garden of the Doom Dooma Company that same day, the situation had turned even more "menacing." While the DC, accompanied with a subedar and a platoon of 25 Assam Rifles marched into the estate, the workers were found demanding "a special issue of rice" from the manager.[6] The pattern continued. Two days later, workers of the Hukanguri garden of the Assam Frontier Company raided the weekly *haat* and *kaya* shops and allegedly decamped with a "good deal of cloth and rice." The same evening, at the Daisijan garden, the manager was assaulted and the garden *babu* "almost beaten to death." The officiating commissioner however commented that on

---

[2] Vide, letter no. 471-F, dated October 30, 1920 from the Officiating Commissioner, Assam Valley Districts, to the Chief Secretary to the Chief Commissioner, Assam, IOR/L/PJ/6/162, 1921.
[3] Ibid.; these incidents were also reported by the Assam Labour Enquiry Committee in their official report published in 1922, see *Report of the Assam Labour Enquiry Committee 1921–22* (hereafter ALEC) (Shillong: Government Press, 1922), especially chapter II.
[4] Vide, letter no. 471-F, dated October 30, 1920, IOR/L/PJ/6/162, 1921, p.1. [5] Ibid.
[6] Ibid., p. 2.

inspecting this estate, Mr. O'Callaghan of the Indian Police found the last information "greatly exaggerated."[7]

From the neighboring North Lakhimpur and Sibsagar districts, similar reports followed. On September 12, 1920 around five hundred workers from the Pathalipam estate walked to the district headquarters of North Lakhimpur complaining that they had not received adequate pay or food. Upon inquiry by the Sub-Divisional Officer, it was discovered that the rice supplied to workers from the garden store was "of very poor quality."[8] The Magistrate also found that due to "finer plucking" and the large number of laborers employed on the estate, the quantum of *ticca* (or additional piecework) earnings had substantially reduced as compared to the previous year.[9] Further outbreaks, categorized in official reports as "minor disturbances," were reported from Sibsagar on September 30 and October 7, 1920. Of course, these instances are not exhaustive and other walkouts and protests by laborers in the Assam valley gardens continued throughout the year. In hindsight, these incidents were a chilling premonition of the more historically well-documented Chargola labor "exodus" in the southern Surma valley in 1921.[10]

Comprising the Cachar and Sylhet districts, tea production in the Surma valley had been traditionally geared more towards volume than quality. Given the postwar crises discussed above, planters in the province resorted to finer plucking and restricted outturn to override surplus stocks and losses. Retrenchment and reduced

---

[7] Ibid., p. 3.
[8] Vide, letter no. 471-F, dated October 30, 1920, IOR/L/PJ/6/162, 1921, p. 5. [9] Ibid.
[10] This section has been compiled from official reports on the Chandpur incident, vide "Report by Sir Henry Wheeler, KCSI, KCIE, ICS, Member of the Executive Council, Government of Bengal with reference to his recent visit to Chandpur," in IOR/L/I&O/ 1277, 1921; "Communiqué by the Government of Assam regarding recent developments in Sylhet," in IOR/L/I&O/358, 1921; "Report of disturbance in Solgai," IOR/L/PJ/ 6/1802, file 2293; "The Assam Tea Garden Riots, Reports, and Extracts," IOR/L/PJ/6/ 1706, file 6733, and *Report of the Assam Labour Enquiry Committee 1921–22*. Opposing viewpoints vis-à-vis the government's position are from *The Amrita Bazar Patrika*, May 20, 1921–July 28, 1921, SAMP MF 7102, Center for Research Libraries (CRL), Chicago, and the Non-Official Enquiry Committee Report on the Chandpur Gurkha Outrage, 1921. For a fuller analysis of the details of the Chargola exodus, see Kalyan Sircar, "Coolie Exodus from Assam's Chargola Valley, 1921: An Analytical Study," *Economic and Political Weekly*, Vol. 22, No. 5 (Jan. 31, 1987): 184–193; Amalendu Guha, *Planter Raj to Swaraj: Freedom Struggle and Electoral Politics in Assam, 1826–1947* (New Delhi: ICHR, 1977); also see Nitin Varma, *Producing Tea Coolies? Work, Life and Protest in the Colonial Tea Plantations of Assam, 1830s–1920s*, Unpublished D.Phil. Dissertation (Berlin: Humboldt University, 2011), and Rana Partap Behal, *One Hundred Years of Servitude: Political Economy of Tea Plantations in Colonial Assam* (New Delhi: Tulika Books, 2014).

labor earnings across gardens followed. Amidst these trying circumstances, laborers in the valley were compelled to buy rice at market rates of 6 rupees per maund, and opportunities for alternative income limited.[11] The DC of Sylhet later admitted that the average tea garden worker was living in indigence "or just above the margin of starvation."[12] By the end of February that year, the situation had turned desperate and the Calcutta-based *Statesman* commented on "coolies going about begging for work at some gardens."[13] Three months later, on May 1 and 2, 1921 two meetings were allegedly held by traveling "non-cooperators" at the Ratabari garden in the Chargola valley of the Sylhet district. Ostensibly organized to discuss the ongoing Khilafat agitation and Gandhi's civil disobedience program, police records claimed that one Radha Krishna Pande of Silchar urged tea workers from the Anipur tea estate to demand pay increases averaging 8 annas a day for men, 6 annas for women, and 3 annas 6 pies for children or walk out. Pande was also reported to have compared tea managers to Satan.[14]

In response, around seven hundred and fifty workers of the Anipur garden marched towards Karimganj town the following day. As news spread, more walkouts commenced across the valley. Laborers from Longai, Adamtila, Lalkhira, Baitakhal, and Eraligul estates joined in. In official records, total numbers of the "exodus" varied between seven and ten thousand.[15] Determined to return to their villages, tea laborers boarded trains bound for Chandpur via Goalando and onward to Calcutta. Faced with an initial group of eighteen hundred souls at Chandpur, and fearing an outbreak of cholera, the local Sub-Divisional Officer (hereafter SDO) arranged for their transport via steamer to Goalando. The Bengal administration (hereafter GOB) then stationed at Darjeeling however disapproved of this repatriation at government cost and expressly forbade further dispatches. As laborers began to stream in Chandpur, the GOB effectively washed its hands of the matter urging tea companies (and their labor associations) to manage the crisis. The District Magistrate of Goalando was also instructed to persuade arriving laborers to return to their respective

---

[11] See *ALEC 1921–22*, Chapter II, and Sirkar, p. 184.
[12] Vide, letter IOR/L/PJ/6/5598 with file I&O, no. 57/21, dated September 12, 1921, "Report from the Government of Assam as to the Adequacy of Wages Paid on Tea Gardens," Asian and African Studies, British Library, London.
[13] Quoted in Sirkar, p. 184.   [14] See *ALEC 1921–22*, p. 10.
[15] Vide, "The Assam Tea Garden Riots, Reports, and Extracts," IOR/L/PJ/6/1706, file 6733; also, Sirkar, p. 185.

gardens. On May 19, 1921, Chandpur town was teeming with around three thousand starving and famished tea workers hoping for their return home. As the Magistrate, the Divisional Commissioner, the Superintendent of Police and the representative of the Tea District Labor Association arrived at Chandpur, several cholera deaths amongst stranded laborers were reported. Several hundred took shelter at Chandpur railway station some distance from the main town center. Meanwhile, fearing incendiary propaganda by the "non-cooperators," the district authorities clamped section 144 of the Indian Penal Code restricting public meetings around seven miles of some tea gardens. The logic reverted to an earlier trope: "outsiders" with narrow political ends had tricked the "hapless coolies" to leave their "contended" life on the tea estates. By the evening of May 19, Chandpur was on the edge. The local Congress leader, Hardayal Nag was also roped in to persuade the workers to leave.

Resolved to move on, batches of workers made repeated attempts to board steamers jettisoned beside Chandpur railway yard. On their third attempt on May 19, around three hundred forty souls were reported to have boarded crowded vessels. The SDO and Mr. J. McPherson, the labor association representative, allegedly asked for the gangway to be forcibly removed resulting in the deaths of several workers. As the situation continued, and further attempts to storm anchored steamers made, emotions ran high on all sides. The following morning, the Divisional Commissioner, Mr. K. C. De wired for military reinforcements of 50 men from the Eastern Frontier Rifles. Amidst fears of further cholera outbreaks, renewed attempts to forcibly board steamers, interference with railway movement, and general "disorder," Mr. De in consultation with McPherson and other junior officers decided to relocate "squatting" laborers from the Chandpur station to a nearby football field. Proposals for arranging shelter for these itinerant workers, along with an agreed donation of 2000 rupees by the labor association were also made. As it is, the presence of local Congress leaders in the town discomfited an already edgy local administration. A forcible removal of laborers from the Chandpur railway station, if needed, had reportedly been decided. Around 10.30 PM on May 20, 1921, the requested riflemen arrived and were marched onto the railway station. Warnings were issued to laborers to leave, but went largely unheeded. In the ensuing melee, Gurkha soldiers were allegedly set upon the workers and a large number of men, women, and children were injured. While the exact number of

Figure 6.2 The junction of the rivers [railway construction scene at junction of Ganges and Brahmaputra rivers near Goalundo], c. 1870 © The British Library Board, Photo 230(44)

laborers wounded, and the magnitude of force employed, remain contested to this day, this miscalculation by the district administration was the proverbial last straw in this entire episode.

As news of the "Gurkha Outrage"[16] rippled across eastern India and beyond, several others joined in the protests. The Congress leader, Bipin Chandra Pal condemned the GOB for walking out on labor problems by stopping free passage.[17] The Indian Mining Federation offered employment to stranded workers in the coalfields, and on May 24, employees of the Assam Bengal Railway struck work in support. Many others, including well-known Bengal Congress leader C. R. Das and missionary Reverend C. F. Andrews actively participated in relief and rehabilitation efforts. Among other things, the Chandpur incident launched the GOB, the planters, and the nationalists on a warpath regarding culpability and cause. These wrangling aside, the events of May 19–20, 1921 catapulted the tea laborer plight – especially those in the Surma Valley – into national and international spotlight.

Politics or Economics?

The aim here is not to provide a descriptive catalogue of labor unrest during 1920–21. Protests, desertion, all-out violence, absenteeism, and labor complaints against overbearing work schedules and managers had been a recurring hallmark of the Assam plantation system since its inception.[18] What made these walkouts and riots of the early twentieth century different? Did labor demands regarding food supply and wages illuminate anything more than the known conditions of work in these estates? Some answers were predictable.

For one, it was widely recognized that "the coolie has had to content himself with a lower standard of living than before the war."[19] As a case in point, consider the following figures for the comparative appreciation in the "standard cost of living" versus the "actual average cash wage" of working men and women in the Empire of India and Ceylon Tea Company Limited.[20]

---

[16] Reported by the Non-Official Enquiry Committee Report on the Chandpur Gurkha Outrage, in *The Amrita Bazar Patrika*, May 31, 1921.

[17] Bipin Chandra Pal, "Tea Labour Exodus and the Government," in *The Amrita Bazar Patrika*, May 27, 1921.

[18] On this, see Muhammad Abu B. Siddique, *Evolution of Land Grants and Labour Policy of Government: The Growth of the Tea Industry in Assam 1834–1940* (New Delhi: South Asian Publishers, 1990), Behal, *One Hundred Years of Servitude*, and Guha, *Planter Raj to Swaraj*.

[19] Vide, letter no. 471-F, dated October 30, 1920, IOR/L/PJ/6/162, 1921, p. 6.

[20] Vide IOR/L/I&O/1449, p. 7, Asian and African Studies, British Library, London.

Plant and Politics

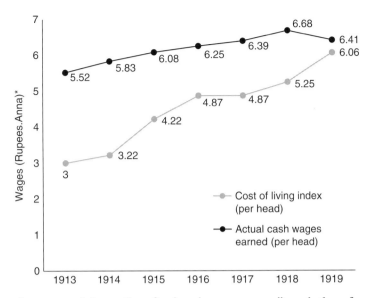

Though prepared by a "professional accountant," and therefore touted as "reliable" by the plantation management, the above figures need to be read with caution.[21] For instance, the data do not make clear the parameters that went into calculating the cost of living index. They also do not indicate whether the cash wage earned applied to *working* labor or to the entire workforce present on the Empire Company gardens. Finally, these averages flatten the wage differential between men, women, and children that varied greatly from one estate to the next depending on nature of work, total workers, earnings from secondary income (paddy cultivation, for instance), and overtime cash allowances. The point, however, is that, even with all these variables, the two indicators above show that whereas the cash wage earned per head remained steady between the prewar and the postwar years (at 6 rupees and 1 annas), the cost of living had risen by almost 100 percent during the same period.

The officiating commissioner was therefore candid in his assessment that unrest amongst tea workers should be no matter for surprise.[22] A successful strike – and resultant pay increase – by workmen of the neighboring Dibru-Sadiya Railway was also seen

---

[21] Ibid.
[22] To put matters in perspective, it was officially recommended in 1906 that lower than statutory wages in the Assam plantations compared with the rest of India was a deterrent to labor recruitment; see *Report of the Assam Labour Enquiry Committee 1906* (Calcutta: Superintendent of Government Printing, 1907), especially Chapter XI. Incidentally,

as contributing to these protests. It was a vicious cycle, or so the narrative ran. The depressed state of the tea industry after the war, overproduction and surplus stocks, and the insistence of quality over volume meant that not all labor could be engaged – and therefore compensated – as before. The commissioner maintained that in comparison with September 1919, for instance, the average wage of Dhoedaam and Pabhajan estates were substantially lower in 1920. To complicate matters further, Marwari *kayas* supplying rice to tea laborers frequently peddled grain of very inferior quality in hopes of making up their own losses. The commissioner reasoned:

> Garden Managers generally, I believe, recognise that an increase of wages would be desirable, but unfortunately at the present moment tea is being produced on most gardens at a heavy loss; few, if any, estates are working at a profit and proprietors and managers are bound to work as economically as they can.[23]

The above report effectively dismissed "outside incitement" and political "contamination" by Gandhian noncooperators as credible causes of labor unrest. Of the latter, the commissioner found very little evidence. The British press, however, alleged that official secretiveness regarding the Assam "riots" had allowed rumors to exaggerate actuality, and fear and conjecture to prevail.[24] The *Morning Post* was quick to suggest that extraneous influence rather than pay was responsible "because ... the conditions under which the natives work in these gardens are far superior to those which prevail elsewhere in that part of the world."[25]

As this trope of political "infection" continued over the next two years, confidential official correspondence continued to repeat the commissioner's early wisdom. In his weekly telegram to the Secretary of State, Viceroy Chelmsford was categorical that high prices coupled with reduction in overtime earnings were the likely reasons for labor trouble in the Assam valley.[26] The Chief Commissioner of Assam however reiterated to the Government of India (hereafter GOI) that unsatisfactory state of the tea industry

---

a proposed offer in 1918 to raise wages in the Doom Dooma district was vetoed by tea authorities.

[23] Ibid., p. 7.

[24] See "The Wrong Way," *Pall Mall*, October 4 1920, vide, IOR/L/PJ/6/6733, Asian and African Studies, British Library, London.

[25] See "Indian Tea Coolies' Wages," The *Morning Post*, October 6, 1920 vide IOR/L/PJ/6/7203, file no. 4340, Asian and African Studies, British Library, London.

[26] Telegram from Viceroy (Home Department) to Secretary of State, dated October 16, 1920, in IOR/L/PJ/6/7203, file no. 4340.

notwithstanding, enabling a laborer and his family to earn a living wage in return for a fair day's work was the first charge upon the current income, of the accumulated capital, of [a] Company.[27]

The Chargola walkout received much more attention. Widely picked up in the nationalist and international press, the plight of the Surma valley laborer and the horrors of the Gurkha charge struck a raw political nerve. Historically speaking, the exact nature, stakes, and ramifications of the Chandpur incident continue to be debated to this day.[28] It is not my intention here to repeat them in toto. Some patterns of contemporary opinion and counter-charges are however relevant to our story. The London board of the powerful ITA claimed that malicious noncooperators had made "pawns" of the tea garden coolies.[29] Indeed, H. M. Haywood, Secretary of the ITA argued that the cause of the Chargola walkout was noneconomic and that agitators had stirred unrest amongst the "wholly unreasonable" workers by exploiting their ignorance and religious superstitions.[30] Haywood regurgitated old managerial wisdom that the laborers were financially better off in the gardens than in their home districts. Sir Wheeler's *Report*, based on his tour of Chandpur from May 29 till June 3, 1921 was decidedly more measured, though he too squarely blamed mismanagement and disorder for the turn of events.[31] Wheeler commented that, given the situation, the "coolies would not have moved had some force not been applied to them."[32] In his assessment, the wounds received were "not excessive" and counter-charges of bayonet use were "without foundation."[33] More crucially, Sir Wheeler was convinced that the *hartals* and strikes were instigated by political aims and were a camouflage for direct action. But the point in his

---

[27] Vide, letter no. 11264, dated November 27, 1920 from the Chief Secretary to the Chief Commissioner of Assam to the Secretary to the GOI (Home), IOR/L/PJ/6/162, 1921, Asian and African Studies, British Library, London.

[28] In addition to the works mentioned under footnote 10 above, also see Rakhahari Chatterjee, "C. R. Das and the Chandpur Strikes of 1921," *Bengal Past and Present*, Vol. XCIII, Parts II & III, Serial No. 176 & 177 (September–December 1974): 181–196; J. H. Broomfield, *Elite Conflict in a Plural Society: Twentieth Century Bengal* (Berkeley, CA: University of California Press, 1968); R. C. Majumdar, *History of the Freedom Movement in India*, Vol. III (Bombay: K. L. Mukhopadhyay, 1963); and Benarsidas Chaturvedi and Marjorie Sykes, *Charles Freer Andrews: A Narrative* (New Delhi: Govt. of India Publications Division, 1971), especially chapter XII.

[29] Circular of the Indian Tea Association (London), July 15, 1921 vide, IOR/L/I&O/1277, Asian and African Studies, British Library, London.

[30] H. M. Haywood, "Tea Labour in Assam and Surma Valleys," letter to the editor, *Indian Planters' Gazette*, May 27, 1921 reprinted in the *Indian Planters' Gazette*, June 4, 1921, accessed at the National Agricultural Library, Beltsville, Maryland.

[31] Vide "Report by Sir Henry Wheeler, KCSI, KCIE, ICS, Member of the Executive Council, Government of Bengal with reference to his recent visit to Chandpur," in IOR/L/I&O/1277, 1921, pp. 1–12.

[32] Ibid., p. 9.   [33] Ibid.

Report that stuck out most was the question of Government "neutrality."[34] He refuted Reverend Andrews' call for GOB's intervention in relief and repatriation operations.[35] For Wheeler, it was improvident for the government to be either party to or judge the merits of labor disputes where it was not directly involved. This idea of government non-involvement was however paradoxical and disingenuous. Having long placated Assam planters with a slew of legislations aiding labor recruitment, Sir Wheeler could have engineered a better alibi than official "neutrality" as his shield. For, as we see in the next section, the 1920–21 walkouts unearthed deep illegalities and planter contraventions of the labor laws that the GOB had over the years instituted. Wheeler also refuted allegations of starvation wages by invoking "eye-witness" worker testimonies to the contrary.[36] Despite acknowledging "coolie hardships" and reduction in postwar work, Sir Wheeler dismissed endemic fiscal causes as alone responsible for the exodus. For him, it had an unmistakable political tinge.[37]

This version found much support from the ITA, the planter lobby and a section of the British press. The Assam Government also issued a formal *Communiqué* on June 6, 1921 regarding its position on the walkout.[38] Aside from brief comments on the "hard hit" gardens of the Surma Valley, and the loss of *ticca* earnings for coolies, this announcement repudiated the "allegation of very low wages" as false. It did not provide further evidence of this claim except to suggest that an investigation into specific cases would be welcome. The *Communiqué* agreed with employers that workers would not have left in such large numbers if "unscrupulous" persons had not filled their minds with wild suggestions. It was under no doubt that "[coolies] were obeying an appeal made to them on other than economic grounds."[39] But opinions – even from within the Assam government – were rarely uniform on this issue. In an internal

---

[34] Ibid., pp. 9–10.
[35] See "Recruited Labour: Sir Wheeler's Interview with Mr. Andrews," Associated Press of India, Chandpur in *The Amrita Bazar Patrika*, June 3, 1921.
[36] Ibid., p. 11.
[37] Years later, the historian J. H. Broomfield made a similar assessment by drawing a neat causality between Congress involvement and the Chandpur walkout: "in the first half of April, 1921, a group of Congressmen from Calcutta moved into the tea garden areas of Assam and started an agitation for higher wages among the tea coolies," in *Elite Conflict in a Plural Society*, p. 215. Broomfield's simplistic argument, and analytic disregard for the deeper connections of the walkout with labor wages and conditions of work were critiqued by later generation of scholars; see, for example, Sircar, "Coolie Exodus from Assam's Chargola Valley, 1921," Guha, *Planter Raj to Swaraj*, and Chatterjee, "C. R. Das and the Chandpur Strikes" for analyses of this debate.
[38] See Indian News Agency Telegram, No. 3, Calcutta, June 6, 1921 vide, IOR/L/I&O/442 (Home/Political), no. 1015, Asian and African Studies, British Library, London.
[39] Ibid.

memo of July 19, 1921 the DC of Sylhet contravened the *Communiqué*'s version by commenting that though dissatisfaction was sometimes due to "agitator's presence," the tea industry will have no answer to criticisms "unless it can show that the labourer is paid a fair economic wage."[40]

The nationalist press's response to the Wheeler *Report* and the Assam Government *Communiqué* was decidedly scathing. *The Amrita Bazar Patrika* (hereafter *ABP*) argued that both were special pleadings on the planter's behalf – the former masquerading as justice, and the latter unabashed in its love for the tea manager.[41] The *ABP* column highlighted the contradictions of both the reports in first admitting – and then backtracking – on the connection between the depressed state of the industry, lower wages, and labor unrest. Satirizing GOB's purported "innocence" throughout the whole episode, the *ABP* editorial suggested that the Wheeler *Report* made the "gall of servitude" clear to the people of India. It further commented that the Chandpur strikes, and the GOB's inaction "brought home to the Indians [that] administration and exploitation are the two hand-maidens of British rule in [the country]."[42] Even the liberal-leaning *Northern Whig* commented that tea employers should have taken advantage of the "fat years" to raise labor wages in the past.[43]

As is evident, public discourse on – and government response to – the 1920–21 labor walkouts in the Assam plantations quickly devolved into a tautological blame-game of culpability and cause. For some, these protests were spontaneous outbursts of labor disenchantment and anger at subsistence earnings. For others, these "disturbances" were opportune alibis to overthrow the British Empire. Spearheaded by Gandhi's call for noncooperation, many saw in these stirrings the dawn of a "national" labor consciousness, and the rise of an "incipient class militancy."[44] Historically, of course, Gandhi's direct involvement in these events remains contested and ambiguous. Though he was critical of planter "tyranny" and provided moral support to the walkouts and to Andrews' efforts at repatriation, Gandhi claimed that he had no knowledge of the Assam labor problem, and did not recommend "a single coolie

---

[40] See IOR/L/PJ/6/5598, dated August 11, 1921, Asian and African Studies, British Library, London.
[41] "Sir Henry Wheeler's Report: A Study in White Washing," *The Amrita Bazar Patrika*, June 5, 1921.
[42] "The Two Documents," *The Amrita Bazar Patrika*, June 9, 1921.
[43] "Indian Tea Gardens: Report on Labour Conditions," *The Northern Whig*, October 19, 1921, vide IOR/L/I&O/1449, Asian and African Studies, British Library, London.
[44] Guha, *Planter Raj to Swaraj*, Second Edition, p. 108.

[there] to strike."[45] He was also vague in his opinion that noncooperation was not against capital or capitalists, but against the Government as a system.[46] As it is, some saw in the Chandpur incident an opportune moment for the Bengal Congress to discredit the shaky indifference of the GOB.[47] For some commentators, of course, these strikes brought home the memory that *Uncle Tom's Cabin* was still alive in the northeastern corner of British India. Sometimes, the government's action paradoxically boomeranged. Entrusted with an official "fact-finding" mission to understand these labor unrests, members of the 1921–22 Assam Labour Enquiry Commission (hereafter ALEC) were told by some planters that their very presence in the gardens led to these labor disturbances.[48]

While none of these explanations are incorrect, or unimportant, I argue that the accountability-versus-origins debate is historically unhelpful. In other words, the anatomy of these charges and counter-charges does not elaborate *how* wages, work conditions, labor impoverishment, and plantation life were mutually implicated in, and in turn impacted, the coming of the 1920–21 crises. It does not clarify what the *Daily Telegraph* meant by "methods of personal management" quoted above, and why the question of political "infection" – or government reaction to the matter – necessarily centered on the question of wages. The dots are left inadequately connected, for the answers lay elsewhere.

A peculiar culture of commerce in the Assam plantations – centered on law, crop agronomics, and work structure – provide an alternate analytic framework to understand these walkouts. Put differently, what made the 1920–21 protests thorny and historically unique was not just its scale and political timing. They highlighted

---

[45] See M. K. Gandhi, "The Lesson of Assam," in *Young India*, June 15, 1921 reprinted in *The Collected Works of Mahatma Gandhi*, Vol. XX, April–August 1921 (Delhi: Govt. of India Publications Division), p. 228.

[46] Just one month after the Chandpur episode, Gandhi wrote in *Young India*: "In India we want no political strikes. We are not yet instructed for them. Not to have political strikes is to forward the cause of freedom. We do not need an atmosphere of unsettled unrest ... we seek not to destroy capital or capitalists, but to regulate the relations between capital and labour. We want to harness capital to our side," in *Young India*, June 15, 1921, p. 228. Gandhi did not clarify what he meant by "regulating" this relation between master and men.

[47] See Chatterjee, "C. R. Das and the Chandpur Strikes of 1921."

[48] See *Report of the Assam Labour Enquiry Committee 1921–22* (Shillong: Government Press, 1922), pp. 4–12; also see the deposition of Mr. A. Moffat, Superintendent of the Jhanzie Tea Company, February 27, 1922 in *Evidence Recorded by the Assam Labour Enquiry Committee* (Shillong: Assam Government Press, 1922), pp. 166–168.

unseen and illegal logics, institutional practices, and estate politics that converged in the making of labor life – and managerial perceptions of "fair" wages – in these plantations.

Culture of Commerce

> The true criterion of the tea garden coolie's wages is the opportunity to earn[49]

Reassessments of the 1920–21 protests have to start with the tea plant – indeed agrarian economics – in the province. The wartime operation of the Assam tea industry was a bubble of sorts. Fueled by a guaranteed two-thirds outturn purchase by the Government Food Controller, and rising Continental demands, tea production reached a staggering figure of 245,385,920 pounds around 1914–15.[50] But these unrealistically happy times soon changed. Demands, especially from Russia drastically shrunk after the war, and the Food Controller scheme was withdrawn.[51] With a vast surplus stock rotting in the London warehouses, the total outturn of black tea in 1920 touched only around 180,332,192 lbs.[52] It was estimated by the London *Times* that close to 150 million pounds of tea was lying unsold and chided the industry for its reckless wartime expansion program.[53] Of course, planters had already warned as early as 1918 that supplies had started accumulating (see Table 6.1 and 6.2 below).[54]

In the postwar period, the tea enterprise came face to face with several operational shibboleths: for one, quantity was not an automatic guarantor of profits. But it had other peculiarities to contend with. Unlike mills and factories, for instance, plantations could not be entirely shut down. As the ITA claimed, doing so would entail a total abandonment of capital sunk. But the industry's most valuable and irreplaceable asset – human labor – could not be dispensed with either. Allegedly costing in the "neighbourhood of

---

[49] Confidential letter from DC, Nowgong, Assam to the Commissioner, Assam Valley Districts, no. 121M dated April 11, 1921, IOR/L/I&O/1449, Asian and African Studies, British Library, London.
[50] See *Report on Tea Culture in Assam for the Year 1915* (Shillong: Government Press Assam, 1916), IOR/V/24/4280–81, Asian and African Studies, British Library, London.
[51] Also, a steep rise in the rate of exchange made trade in rupees unfavorable for Indian tea companies; see appendix VI of *Report of the Assam Labour Enquiry Committee 1921–22* for details.
[52] See *Report on Tea Culture in Assam for the Year 1920* (Shillong: Government Press Assam, 1921), IOR/V/24/4280–81, Asian and African Studies, British Library, London.
[53] Quoted in Sircar, "Coolie Exodus from Assam's Chargola Valley," p. 184.
[54] See *The Indian Planters' Gazette*, September 14, 1918, accessed at the National Agricultural Library Manuscript Collections, Beltsville, Maryland.

Table 6.1 *Outturn of manufactured tea (in lbs.) per acre*[55]

District	1916	1917	1918	1919	1920	1921	1922
Cachar	576	591	543	515	536	371	468
Sylhet	590	602	591	519	561	367	443
Darrang	720	667	701	648	671	544	496
Nowgong	615	615	587	570	554	433	462
Sibsagar	665	658	735	630	555	513	512
Lakhimpur	788	745	751	762	699	569	640
Assam total average:	664	651	664	613	597	434	513

Table 6.2 *Tea produced, and corresponding prices, for twelve months ending March 31*[56]

	Brahmaputra Valley			Surma Valley		
Year	Number of Packages	Price per lb.		Number of Packages	Price per lb.	
		[Anna	paisa]		[Anna	paisa]
1919	267,816	9	1	140,275	6	9
1920	244,232	8	4	177,467	7	8
1921	244,280	6	3	196,025	3	7

Rupees 150 per head" to recruit by one estimate,[57] they were "precious investments."[58] Of course, as pointed out in chapter 4, incidentals needed to keep labor in survivable capacity – including health, hygiene, nutrition, and compensation – could be tinkered with, and came towards the very end of planter concerns. Indeed, the ITA was unabashed in arguing: "owing to the present financial crisis, schemes for improving sanitation, etc. must remain in abeyance."[59] Lastly, the enterprise had law to deal with. Labor legislations in the Assam plantations were necessary evils as far as managers were

---

[55] Compiled from *Report on Tea Culture in Assam for the Year 1920–21*, IOR/V/24/4280–81, Asian and African Studies, British Library, London.
[56] Ibid., p. 3; 1 rupee was approximately 16 annas, while 4 paisa equaled 1 anna.
[57] J. C. Higgins, DC Nowgong, Assam in letter no. 121M dated April 11, 1921, IOR/L/I&O/1449, Asian and African Studies, British Library, London.
[58] Ibid.
[59] See *Detailed Report of the General Committee of the Indian Tea Association for the Year 1920*, Mss Eur. F 174/611, p. iv, Asian and African Studies, British Library, London.

concerned. If only in theory, they regulated contractual obligations, hours of work, and pay. To be sure, all these stipulations had long been sidestepped or manipulated in these estates; the 1920–21 fiscal crisis was not a catalyst in this regard.

In the midst of these institutional strangleholds, the tea industry devised a two-pronged strategy to tide over mounting losses. One was to target the crop, and the other to engineer a differential wage structure. Let us look at the first policy in some detail. Around September 1919, the London board of the ITA cabled its Indian counterparts to embark on a vigorous "Crop Restriction Program" (hereafter CRP) during the 1920 season.[60] Pragmatically speaking, this meant an emphasis on tea quality over volume. The only practical way to do this was to select the finest, albeit limited cache of leaves during the harvest season. Even this came with attendant problems – human and geographical. While the northern Assam valley gardens regularly produced leaves of this kind, the southern Surma valley had made its mark on volume. It was naturally hit hard during the implementation of this scheme. Also, fine plucking needed skill and expertise and could therefore engage only some of the workforce previously recruited during the prewar boom. The program also came with additional caveats.

The CRP guidelines recommended that the crop output be reduced by up to 90 percent of the average of the preceding five years.[61] Alternately, companies were asked to cease plucking by November 15 – a good four months before the traditional end of harvests. Even Ceylon (now Sri Lanka) was asked to cooperate.[62] It was felt that terminating the season around November would automatically take care of new gardens with immature saplings, large holdings with vast, unremunerative extensions, and estates with low per acre yield. Mostly in place by late 1920, these measures reduced tea production by around 5 million lbs. compared to the previous year.[63] Close to 11,191 acres of cultivation was abandoned.[64] Additionally, the ITA recommended that crop shipments to London be restricted by 5 percent of the total outturn in

---

[60] The same *Times* review mentioned above noted ITA's crop reduction program with some satisfaction, and suggested that "sanity" had finally returned on the Assam tea estates, see *The Times*, January 28, 1921 quoted in Sircar, "Coolie Exodus from Assam's Chargola Valley," p. 184.
[61] See *Detailed Report of the General Committee of the Indian Tea Association for the Year 1920*, p. 4.
[62] Ibid., p. 5.
[63] See *Report on Tea Culture in Assam for the Year 1920*, IOR/V/24/4280–81, p.2.
[64] *Report on Tea Culture in Assam for the Year 1921*, p. 1.

each of the months of October, November and December 1920.[65] Delayed exports meant that manufactured tea, sitting on estates over long periods and especially the monsoons, arrived dull and soggy. For instance, the 1919 yield had reportedly not been dispatched till June 1920.[66] It was also suggested that liners agree to disperse imports to outlying ports of London so as not to overcrowd already crammed warehouses in the city. But one company's crisis was another's liability. The 1920 Liners Conference disagreed with these orders in toto and the reduction was raised by 8 percent.[67] Alongside, the Government of India was requested to remit the export duty of 1 rupee and 8 annas per 100 lbs. of tea that had been levied during wartime in 1916. Of course, the Income Tax Act VII of 1918 had already riled the industry, as discussed in Chapter 2.

However, agro-economic decisions are rarely self-solving. The direct burden of the CRP was borne by institutional (and non-formal) manipulation of labor wages. This was a complex and invidious operation. Wage tampering did not simply devolve around *what* workers earned in these estates. Under a peculiar managerial regime, it involved regulating *who* could earn in the first place, and *how much*. As pointed out, under the CRP and finer plucking, fortunes had already turned heavily against the laborer's favor. These policies only added injury to the wound. There was also a legal issue involved: if law fixed monthly wages and hours of work, how could planters institute variable incomes?

To get to the bottom of the wage question, the relationship between legal stipulations and plantation practice needs to be examined. In theory – indeed in law – the Assam Act laborer was guaranteed a "minimum" monthly pay.[68] Act VI of 1865 was the first to institute this regulation whereby 5 rupees for men, 4 rupees for women, and 3 rupees for children under contract were recommended.[69] However, in praxis, these were mostly followed in the breach as planters interpreted these numbers as minimum *rates* and not *fixed* in principle. In other words, managerial circumvention of legal wisdom categorized these

---

[65] See *Detailed Report of the General Committee of the Indian Tea Association for the Year 1920*, p. 6.
[66] See appendix VI, *Report of the Assam Labour Enquiry Committee 1921–22*, p. 116.
[67] Ibid., pp. 6–7.
[68] The difference between "Act" and "Non-Act" labor in Assam has been discussed in Chapter 4. In addition to these two categories, workers were also recruited from contiguous village settlements known as "basti" or *faltu* labor. This chapter does not focus on this group in elaborating arguments regarding wages. See Behal, *One Hundred Years of Servitude*, chapters 3 and 4 for a discussion of basti workers.
[69] See *Report of the Assam Labour Enquiry Committee 1906*, p. 8.

monthly earnings as "contingent" on completing a quantum or volume of work (*hazira*) per head per day. A certain rate (*neerikh*) was then allocated to this task-work which was either paid weekly or monthly – a policy that varied from one estate to the next depending on planter preference, work culture and size of holding. In addition to this task-work, workers could also potentially earn additional income by performing piece-work (*ticca*) beyond regular working hours. Under the weekly payment scheme – also called the "unit system" – payments were made depending on the nature of work completed. For the three most rigorous tasks, namely hoeing, pruning and plucking, rates varied between one anna for the former two and one pice for the latter. Other employees of the estate, including watchmen, *sardars*, accountants and *jamadar* were usually paid monthly.

As is evident, the labor wage policy of the Assam plantations had sidestepped legal stipulations from the very beginning. Aside from tinkering with minor aspects of successive labor laws – and periodically appointing enquiry committees – the GOB did very little throughout this period to remedy managerial disregard of legislative injunctions. Commissioners appointed in 1868 to recommend changes in labor legislations were candid in their assessment that the 1865 Act had failed to secure the interests of the "coolie."[70] To come back to the question of wages, another attempt was made two decades later to "fix" minimum earnings. Act I of 1882 once again attempted to regulate pay: 5 rupees monthly for men, and 4 rupees for women for the first three years of contract, and one rupee higher for the last two.[71] Astoundingly, even this remained a statute-book feature and flagrant manipulation of the task- and piece-work referred to above continued. In fact, the DC of Darrang remarked in 1883 that it was an utter waste to regulate minimum pay as work varied wildly from one garden to the next depending on season, soil quality, leaf produced, and "quality of coolie."[72] Writing around 1888, Luttman Johnson, Commissioner of the Assam Valley districts was quite certain that no amount of inspection in the Assam estates would improve wages.[73] The pattern continued well into the twentieth century.

The above culture of work effectively meant that the monthly (or weekly) earnings of tea laborers depended on everything other than guaranteed income. It was, to put matters plainly, each to his/her own. Bodily fitness, available opportunities, nature of task involved, production policy (quality and/or quantity), plantation size, climate and

---

[70] See *Report of the Commissioners Appointed to Enquire into the State and Prospects of Tea Cultivation in Assam, Cachar and Sylhet* (Calcutta: Calcutta Central Press Company Ltd., 1868).
[71] Vide, *ALEC 1906*, pp. 9–10.
[72] Quoted in Behal, *One Hundred Years of Servitude*, p. 65.   [73] Ibid., p. 66.

184   Tea Environments and Plantation Culture

Figure 6.3 Hoeing between the plants on an Indian tea plantation, 1911, postcard by unnamed artist © Mary Evans Picture Library

hydrology, pests and soil could impact differential earnings. Added to these, difference in opinion between planters, the Calcutta agency houses, the ITA and its London board could also influence wages from one garden to the next. It was an unregulated house in all senses of the term. No *one* index of earnings is therefore suitable for the Assam plantations as a whole. These ghosts of the past came back to haunt the government during the 1920–21 crisis. Urgently seeking wage figures and historical trends to rebut domestic and home criticisms of low labor pay, the GOI realized that statistics was its worst defense. In a confidential report, the Second Secretary of the Assam government replied to the GOI home department that "the production of uniform, clear and convincing statistics [is] impossible [because] of the varying conditions of different localities and from the fact that each garden or group of gardens has evolved for itself the method of remuneration which experience has proved most suitable."[74] What was this method, and what did it mean for workers on the ground?

Consider hoeing and plucking. For the *hazira* system described above, more work should have ideally led to enhanced incomes. But this was hardly the case. For hoeing, usually done by men, a unit of measurement,

[74] Letter No. 5842-F dated July 19, 1921, IOR/L/I&O/1449, Asian and African Studies, British Library, London.

nal (roughly equal to 12/13 feet) was used. Either deep hoeing (usually once a year) or light hoeing (five to seven times annually) were carried out. The average *neerikh* for the former was around 12–16 square *nals* and 25–30 square *nals* for the latter. Though planters estimated that light hoeing took around 3–5 hours, much depended on variables outside the laborer's control. In other words, hardness of soil, terrain, bodily fitness, and state of the garden (clean or forested) impacted completion of the quantum of the daily *neerikh* – and thereby the earning earmarked alongside. Thus, the managerial dictum that tasks ensuring minimum wages could easily be completed within 3–3½ hours,[75] and thereby leaving laborers ample time for *ticca* income, flattened earning differentials into one unitary estimate. These practices, prevalent for decades in the Assam plantations, came under significant scrutiny during the 1920–21 walkouts.

The question of plucking was even more complex. Here, the volume of leaf collected – usually paid at a pice per pound – was intimately tied to the kind of tea produced. As mentioned, the northern Assam valley tea districts (such as Darrang, Lakhimpur, and Sibsagar) produced crop of high quality. Finer plucking was the order of the day in this part of Assam though circumstances varied from one area to the next. Conversely, the southern Surma Valley traditionally relied on quantum – and thereby coarser plucking – to override the kind of tea produced from the region. But these distinctions do not tell the full story. The primary unit of measurement for plucked leaf was weight, not completion of task assigned. In other words, the average *neerikh* assigned for a day's minimum wage was 20–24 pounds of green leaf.[76] But several externals – stalks and waste collected, weight of the basket, and "undesirable leaf" – were deducted from this calculation. Even weather influenced these measurements. Excessive rain and moisture content could directly impact the quantum of crop collected at the day's end and reduced worker income. The 1921–22 ALEC reported that in the southern Surma Valley, deductions of up to 25–50 percent of the total leaf collected was practiced. Thus, women bringing in 24 pounds would receive payment for only around 12 on especially wet days.[77] It is needless to add that the *capacity* to collect was rarely uniform across workers and included numerous other factors such as health (including reproductive health) and skill among others.

The idea of minimum wage – in theory and praxis – meant very little in the Assam gardens. At a North Sylhet district meeting

---

[75] See *Report of the Assam Labour Enquiry Committee 1921–22*, p. 33. [76] Ibid.
[77] Ibid., p. 34; also, Sirkar, "Coolie Exodus from Assam's Chargola Valley," p. 184.

in October 1921, this was expressly spelt out: "we consider it most necessary to pay coolies on a sliding scale giving a lower rate of wages for uncompleted tasks."[78] Purportedly, as the DC quoted at the beginning of this section argued, the "true criterion" of a worker's wages was the unlimited "opportunity to earn."[79] Effectively, guaranteed and common income across the workforce – despite legal stipulations – was unheard of. I argue that this culture of differential wages formed the backbone of worker discontent during the 1920–21 walkouts. These protests highlighted to the world the unseen market – and extra-legal – logics that ran the Assam plantation system. For the industry, of course, it was an expedient balance of payments game.[80] Consider the following earned monthly income for men and women in eight major tea districts between 1913 and 1920:[81]

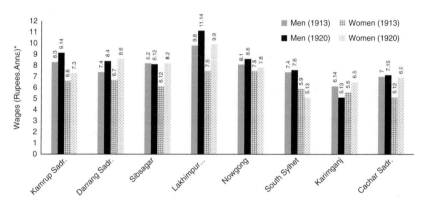

At first glance, these figures seem to fit in with our above-discussed state of plantation affairs. Thus, while earnings in the Brahmaputra Valley remained steady, and marginally increased (especially for women) in 1920, those in the southern Surma Valley (especially for men) either leveled off or decreased during the postwar slump. For South Sylhet, incomes for women in 1920 show a sharp downturn, as is

[78] Ibid., p. 57.
[79] Vide letter no. 121M dated April 11, 1921, IOR/L/I&O/1449, Asian and African Studies, British Library, London.
[80] To be sure, recruitment patterns were closely connected to these wage practices. Having enlisted close to 110,376 souls during the war-bubble, the industry continued to hire indiscriminately in the run-up the 1920–21 slump. Close to 222,171 men, women and children made the journey to Assam in 1918–19; compiled from the Annual Immigration Reports and quoted in *Report of the Assam Labour Enquiry Committee 1921–22*, pp. 103–4. The concept of differential earnings – especially enforced by retrenchments, subsistence wage, and CRP during 1920–21 – fully evidenced itself in the scale and extent of worker discontent of the time.
[81] Vide IOR/L/I&O/1449, Asian and African Studies, British Library, London.

the case in Karimganj for men that year. Despite these seemingly obvious indicators, there are several omissions and caveats to such statistical representation of reality. For one, the data source used does not clarify whether *ticca* earnings were included in monthly returns – a misleading practice widely followed by planters during this period. *The Bombay Chronicle* highlighted this "scandal [of] calculating over-time earnings in total wages" in a scathing report published in 1922.[82] The 1921–22 ALEC also issued a similar statement. Also, wide agronomic and managerial variation from one estate to the next meant that a uniform method of wage assessment and tabulation was meaningless, indeed impossible for the entire plantation system. The DC of Darrang confided to the Commissioner of Assam Valley Division that even comparing wage returns between the hundred or so gardens in his districts – given their "different systems of management and accounting" – was apt to be fallacious.[83]

There was another invidious element to this wage culture that emerged in the 1920–21 aftermath. As criticisms regarding the non-competitive pay of the Assam tea laborer poured in (workers in neighboring oilfields earned 12–15 rupees monthly on an average), the industry rallied back with the idea of "concessions" or subsidiary income. Under this argument, life essentials provided by the management: medical care, housing, clothing, rice supplies, firewood, festival leave, repairs, water supply, and grazing rights formed part of a laborer's overall earnings, and had to be accounted for. In other words, each one of these "free benefits" – material, social, psychological – allegedly carried undeclared monetary equivalents and were plantation sops unavailable to workers elsewhere. Unsurprisingly therefore, companies began to tabulate concession rice as incurred "losses" in counteracting charges of inadequate compensation. Conversely, A. J. Laine, the DC of Darrang, argued that if these "compassionate allowances" were included, *total* labor wages in his district could be seen to have appreciated by at least 35 percent between 1913 and 1920.[84] As we see below, there were deep paradoxes to this line of reasoning. In any case, the facility of ex-garden cultivation was a favorite trope of planters and the tea management in advertising life on these plantations as desirable and "lucrative."[85] To be sure, laborers did take up considerable portions of land within or in the vicinity of these estates. By one estimate, close to 126,951 acres of land

---

[82] See "Planters' Dominion," *The Bombay Chronicle*, November 4, 1922, IOR/L/E/7/1181, no. 57, Asian and African Studies, British Library, London.
[83] Vide letter dated June 7, 1921 in IOR/L/I&O/1449, p. 6.   [84] Ibid.
[85] See *Evidence Recorded by the Assam Labour Enquiry Committee 1921–22.*

had been taken up in 1920–21.[86] From the managerial perspective, this "privilege" needed to be merged into wages – whether actual or promissory. Reality, of course, was far murkier. For one, available agricultural land varied significantly from one province to the next. While the northern Assam valley districts had unacquired wastelands for settlement, the tea lands in Surma valley had pretty much wiped out cultivable property. Secondly, even in theory, laborers could only cultivate outside of their garden time. It was, so to speak, *ticca* earnings. Thirdly, rents were often charged for these holdings ranging from 8 annas a bigha (one-third of an acre) to 2 rupees a bigha. Lastly, as the 1921–22 ALEC eventually diagnosed, cultivating crops for everyday use had become a matter of necessity – not luxury – for resident workers in the face of stationary wages and mounting cost of living.[87] As Gandhi argued: "grazing ground is to [the coolie], almost like breathing, indispensable."[88] In its final report, the ALEC was equivocal that rate of pay in estates could not be kept low under the pretext – and promise – of subsidiary income.[89] It had reason to believe that this had been the practice for many years leading up to the 1920–21 walkouts.

The theory of concessions as additional income was riddled with deep ironies and contradictions. For one, law (especially sections 132–136 of Act VI of 1901) had mandated that planters and companies provide for health, sanitation, housing and rice. But as we have already seen in Chapter 4, regulation and profiteering shared an uneasy relationship in the Assam gardens. With the repeal of Act VI of 1901 in 1908 (and again with Act VIII in 1915), these stipulations effectively ceased. With tea prices falling, and stocks mounting, time was opportune to categorize welfare provisions as concessions. As the 1921–22 ALEC noted, the term itself was of relatively recent origin.[90] To be sure, the ruse of extra plantation privileges had always been a planter favorite in validating stationary wages. The postwar slump was an especially apt time to bring it back in circulation. The paradox, however, remained. If all of these so-called privileges had a cash-value – and therefore a price – how was it to be

---

[86] Compiled from the *Resolution on Immigrant Labour for 1920–21*, Statement X, Assam State Archives, Guwahati.

[87] The DC of Darrang himself mentioned that cost of living in the district had gone up by almost 100 percent between 1913 and 1920; vide IOR/L/I&O/1449. The Bengal Chamber of Commerce computed this figure to be around 139.95 percent; see *Report of the Assam Labour Enquiry Committee 1921–22*, pp. 68–69.

[88] See M. K. Gandhi, "A Planter's Letter," in *Young India*, June 29, 1921 reprinted in *The Collected Works of Mahatma Gandhi*, Vol. XX, April–August 1921 (Delhi: Govt. of India Publications Division), p. 299.

[89] *Report of the Assam Labour Enquiry Committee 1921–22*, p. 71. Indeed, the 1906 ALEC had also issued a similar warning.

[90] See *Report of the Assam Labour Enquiry Committee 1921–22*, p. 21.

passed on to the laborer as income? Here, planters reached an impasse. While theoretical manipulation was easy, its practical application was another matter.

Firstly, it was possible to compute the money equivalents of only some of these "benefits" – rice, for instance. For others (such as housing) variations across estate size, labor strength, location, and production needs meant that standardizing "value" on overheads was an impossible – even unprofitable – solution. The argument had come full circle here, for the burden of disbursing cash equivalents ultimately fell on the employer, or planter in this case. Thus, while it was expedient to suggest – even enumerate – concessions as dormant income, it was best left at that. Pragmatically speaking, of course, laborers could not make live by "eating" healthcare and housing "privileges."

Despite these contradictions, planters and the ITA continued to project extras as unaccounted labor wage. A last moment of irony came with the 1921–22 enquiry hearings. Pressed on the desirability of offering cash in lieu of concessions by ALEC members, managers recommended against the move. This time, another alibi was used as justification. It was felt that the "coolie" was in a state of intellectual infancy and would not "understand" the conversion of these benefits to cash. As the manager of Panitola tea estate deposed before the committee: "the average [laborer] is a child in point of mentality and his thinking has to be done for him."[91] Planters further suggested that inflating "coolie income" with cash for overheads would have an inverse effect on productivity. As the cliché ran, laborers were wont to work only for bare survival. The DC of Darrang was convinced that increased wages led to higher absenteeism and a rise "in the incidence of drunkenness."[92] Paternalistic theories also complemented these views. Managers testified that indirect remuneration brought laborers "in closer contact" with employers, and added a "personal touch" to their relationship. In their eyes, doing away this "necessary" feature of plantation life was absolutely "undesirable."[93]

The *idea* of concessions as extra wages was flawed on all sides. From the manager's perspective, it was a convenient – even necessary – theory in the face of the 1920–21 crisis. They used it time and again to validate staggeringly low earnings of the Assam tea laborer vis-à-vis peers elsewhere. In practice, however, a majority of planters were against its conversion to cash allowances. They were unwilling to standardize and take

---

[91] Quoted in *Report of the Assam Labour Enquiry Committee 1921–22*, p. 29.
[92] DC Darrang to the Commissioner, Assam Valley Districts Division, vide IOR/L/I&O/1449, p. 9.
[93] Cachar Sub-Committee reply to Question 73, cited in *Report of the Assam Labour Enquiry Committee 1921–22*, p. 29.

on this additional burden across the province, and for all workers. Further logics were revealed in the process. It came to light that laborers almost never partook of tea profit "bonuses." Planters and agents argued that their "illiterate" and "infantile" minds would be unable to distinguish between a fat year and a lean one. Speaking for the present crisis, W. A. Cosgrave, DC of Lakhimpur stressed that workers could not be made to understand price fluctuations, and that "non-payment of commission ... when market is dull would probably operate as another potential cause of discontent."[94] The DC of Nowgong (now Nagaon) suggested that such "imaginary grievances" were best rooted out by not introducing them to this concept at all. For him, it was also "inequitable" that *all* workers – irrespective of capacity – be disbursed with company windfalls. Proportional bonuses according to work done was also unworkable in his opinion as it would entail "vast and unnecessary calculations."[95] These reasoning notwithstanding, raising wages was not always in the planter's direct control. The 1920–21 crisis – and subsequent inquiries – highlighted the fraught and uneasy relationship between men on the ground, Calcutta agency houses, and overlords in the London office. In one such instance, it was reported that the proposal to raise labor wages by Mr. Woods of the Doom Dooma Tea Company in 1919 was summarily dismissed by the London board.[96] In fact, even the ITA acknowledged that there was a "weak link" and trust deficit between managers and agents in the running of tea affairs.[97]

In closing our discussion about wage culture, a comment on law – or lawlessness – in the Assam gardens is in place. As with labor health, legal diktats on worker compensation were selectively – and expediently – picked up. While planters had long tampered with the spirit of minimum pay, penal control was a much-desired legislative feature. By the early twentieth century, the provision of arrest without warrant – granted by Act VI of 1901 – had become Assam's notorious managerial hallmark. Ostensibly providing planters "greater hold over his workforce," penal powers was repeatedly blamed for the region's recruitment problems. As this dreaded act lapsed in 1908 (and further amended in 1915), planters and agents summarily invoked the "beneficial features" of another law – Act XIII of 1859. But this a paradox of sorts, for this Act

---

[94] W. A. Cosgrave to the Commissioner, Assam Valley Division, dated April 19, 1921 quoted in *Evidence Recorded by the Assam Labour Enquiry Committee*, p. 112.
[95] Vide letter no. 121M dated April 11, 1921, IOR/L/I&O/1449, p. 19.
[96] "Note on Labour Conditions in Lakhimpur District," in *Evidence Recorded by the Assam Labour Enquiry Committee 1921–22*, p. 109.
[97] See *Detailed Report of the General Committee of the Indian Tea Association for the Year 1921*, p. xi, Mss Eur. F 174/612, Asian and African Studies, British Library, London.

XIII disallowed penal arrests without warrant. In fact, prosecution terminated worker contracts. Why did planters prefer its use? Investigations in the aftermath of the 1920–21 crisis revealed that even Act XIII had been manipulated to the employer's advantage. For one, it came to light that penal contracts continued to be executed. As *The Bombay Chronicle* argued: "it is a case of penal law, in civil contracts, arbitrarily enforced … by capitalists who assume the god, affect to nod."[98] In other words, while Act VI had lapsed in letter, planters continued to use its spirit in Act XIII contracts by association. No explanations were provided as to why this was allowed despite the full knowledge of the Calcutta authorities. Secondly, it was discovered that agreements beyond one year – disallowed by the amending Act XII of 1920 – was flagrantly practiced. The 1921–22 ALEC reported that tea companies often feigned ignorance of this last legislation to take on workers for anywhere between 626 and 939 days. The Assam Company allegedly paid commissions to its clerks and *sardars* on the execution of these illegal contracts.[99]

On their part, planters and agents argued that Act XIII had a "moral effect" in binding laborers to their estates. The everyday reality of this euphemism revealed even more skeletons in the cupboard. As it turned out, an elaborate system of "contract bonuses" had been devised as part of the Act XIII agreements. Under this policy, managers advanced lump sum cash – ranging from 10 to 24 rupees – in lieu of contracts signed. Sometimes, benefits in kind were also provided. But these were not grants of largesse. Though estate policy differed from one to the next, bonuses were eventually "recovered" in the form of wage-deductions and contract extensions in the event of non-payment. In a sense, the cash for contract was a double bind: one the one hand, it gave workers the illusion of wealth; on the other, it tampered with their wages and tied them to a never-ending cycle of managerial debt. Numerous cases of litigation and imprisonment against "non-recoverable" advances emerged during the ALEC inquiries. Unsurprisingly, therefore, the majority of planters opposed the idea of Act XIII's repeal that came up in the wake of these walkouts.

But its "formal" end was already in sight. Despite intense planter objections and ITA lobbying, the GOI finally revoked Act XIII from Assam in 1926. With its demise, the long career of indenture recruitment in the Assam plantations finally came to a close. To that end, the events of 1920–21 were absolutely crucial: they had exposed the extra-market and extra-legal workings of the tea industry beyond doubt.

[98] See *The Bombay Chronicle*, November 4, 1922, IOR/L/E/7/1181, no. 57.
[99] See *Evidence Recorded by the Assam Labour Enquiry Committee 1921–22*, p. 79.

Plant *and* Politics

Around the time of the Chargola crises, a song attributed to the Assam tea laborer appeared in the nationalist press:

> The "Englishman" may have his say
> That we're there very happy and gay,
> But starved we were from day to day,
> And God knows how like dogs we stayed.
> ...
>
> At Chandpur tho' we stranded be;
> Still have joy as we are free
> From planters' clutches, tooth and claw
> And their big rapacious maw
>
> Shout "Gandhi's Jai" and cheer up all,
> March on, brothers, with Freedom's call.[100]

The historical significance – and impact – of the 1920–21 walkouts cannot be underestimated. Not only did they hasten the end of the indenture system in Assam, they also laid the ground for more organized labor union activity to emerge around the mid-1930s. On his part, the Congress leader, C. R. Das, emphatically claimed that the pattern emerging from the 1920 to 1921 strikes – or Chandpur more exclusively – were neither labor-related nor political, but "national."[101] For planters and the Calcutta administration, of course, the question of wages and external "contamination" more than explained the nature and origins of these "disturbances."

But, as I have argued, the 1920–21 walkouts cannot be limited to *a* neatly identifiable cause, nor did its uniqueness emerge from *a* distinct ideological agenda. For one, the crisis – and the government scramble to manage order – did not result from its avowed nationalist intent. Nor were they expressly directed at questions of global labor conditions and worker solidarity. In other words, these protests were not about politics *without* but of policies *within* the tea estates. The apologetics of "Gandhian infection," "national strikes," "Gurkha outrage," and "workers' awakening" – though resonant – deflects attention from the chain of events leading up to and following the labor walkouts. These singular causalities flatten the multiple and unseen plantation logics that led to worker impoverishment and backlash in Assam. As shown, these logics included questions of law, work culture, differential earnings, and crop agronomics.

---

[100] "The Cooly's Song," *The Amrita Bazar Patrika*, June 17, 1921.
[101] See Chatterjee, "C. R. Das and the Chandpur Strikes of 1921," p. 189.

Exegeses of these protests – whether colonial, nationalist or historiographic – have been obsessed with the issue of culpability and cause. In doing so, they have often devolved into a blame-game of instigator and victim. With the Chandpur incident especially, the public discourse – and GOB's response – largely centered on whether Gurkha soldiers needed to be "unleashed" on harmless and hapless tea laborers on the night of May 20, 1921. The subsequent nationalist anger mostly invoked the "scandal" of this uncalled-for government action and its indifference in repatriating wounded workers thereafter. For instance, Rai Radha Charan Pal Bahadur of the Bengal Legislative Council recommended to the GOB that any committee inquiring into the event needed to probe why soldiers were "employed on coolies," identify those responsible for the same, and enact means to prevents such excesses in the future.[102] There were no resolutions drawn to assess the ground realities of the plantations leading up to the crises. A report in *The Amrita Bazar Patrika* of June 24, 1921 similarly lashed out at the "motive and mainspring of the outrage" but said little more. As it stands, the bulk of the correspondence on Chandpur rarely dwelt on the mechanics of tea operation in Surma valley and *why* it forced more than 3000 souls to walk out in the first place. As Reverend Andrews presciently observed from his first-hand experience of the event: "[such] utter misery could not have been reached merely by the march down."[103] The fact that the nationalist press had very little to say about the 1920 labor protests in the northern Assam valley is another matter.

The hermeneutics of causality surrounding the 1920–21 events prove to be self-serving and restrictive. In trying to "fix" and brand its origins, these explanations deviate attention from the endemic extra-legal, agrarian, and plantation structures that drove workers to starvation and unemployment. Consider that one government report blamed "inferior type of coolie recruiting" for the coming of these protests.[104] In this version, workers of recent import, unacclimatized to the local conditions, work culture, and disease environment provided the fertile base for Congress agitators to exploit and misuse. Another suggested that the influenza epidemic of 1918–19 "incapacitated the coolie from doing a fair day's work or earning a living wage" in Assam.[105] In this convenient epidemiological alibi, the cause of these

---

[102] See "Bengal Legislative Council: Some Important Resolution, Committee on Chandpur Incidents," *The Amrita Bazar Patrika*, July 2, 1921.
[103] See "Causes of Exodus: Mr. Andrew's Experience," Associated Press of India, May 27, 1921 in *The Amrita Bazar Patrika*, May 28, 1921.
[104] See *Report of the Assam Labour Enquiry Committee 1921–22*, p. 12.
[105] Quoted in ibid., p. 12.

walkouts shifted from the work regime to an externality beyond the planter's control. Even the wage debate around these events are unhelpful: by mostly focusing on whether laborers were underpaid or not, it sidelined the problem of *how* they were paid, and *who* could effectively earn in the Assam estates. The historiography of labor uprising is also one-dimensional: for not only were these events *not* exclusively about worker solidarity, they were much more than the "product of an interaction between the Gandhian impact on primitive minds and [an] incipient class militancy."[106]

The 1920–21 labor struggles were an important watershed in the history of the Assam tea plantations. But political timing, nationalist ends, and a discourse of causality do not explain *why* it riled the GOI, planters and the ITA throughout this period and beyond. I have argued that what these walkouts exposed was an elaborate network of extra-market, informal, and illegal manipulation of wage structure, crop economics, and work culture that ran the tea enterprise in the province. While this culture of commerce had been in operation for decades, the peculiar crisis of the postwar period magnified its hold over the lives and limbs of workers in the province – and ultimately manifested itself in worker discontentment during this period. The provincial and imperial anxiety to assess, manage, and contain the situation is ample testament that these invidious workings of the Assam plantation system had been laid bare for everyone to see. Ultimately, the 1920–21 crises highlighted that economy, ecology, and labor life were inextricably connected in these gardens of eastern India.

---

[106] See Amalendu Guha, *Planter Raj to Swaraj*, p. 108.

Conclusion

Did the 1920–21 walkouts induce long-term betterment in the working conditions of the Assam tea laborer? Did the demise of the indenture regime in 1926, and the withdrawal of the dreaded 1901 Act signal permanent changes in plantation work culture, wages and labor health? Were the epidemiological, ideological, environmental, legal, and economic debates explored in this book finally settled after the Great War? Were the tea bugs vanquished for good? Inquiries into the tea enterprise's state of affairs in the decades preceding India's independence – as well as its functioning in the present – tell otherwise. They highlight the long persistence, and unsolved entanglement, of these questions in the Assam plantation world.

To be sure, the economic footprint of the tea industry in the province remained steadfast in the interwar periods and beyond. Despite a severe slump in the 1930s, Assam contributed more than 40 percent of India's total tea exports in 1929. The crop alone accounted for close to 8 percent of *total* exports from British India during the same period.[1] And Assam employed more than half of *all* laborers employed on the subcontinent's plantations at the time. This book shows that an early feature that ensured such staggering success, namely the colonial state's agrarian largesse towards private speculators, made for much of the industry's exceptional – and unregulated – status in the region. And it went further than consolidating the economic clout of the so-called Planter's Raj. In addition to run-ins with the forestry department as explored in Chapter 5, these vast tracts of plantation property influenced patterns of peasantization, land settlement and reclamation, disease environments, wildlife movement, legal control, and

---

[1] Cited in *Report of the Royal Commission on Labour in India* (London: His Majesty's Stationery Office, 1931), p. 352.

power-relations between the tea enterprise and the government.[2] For instance, the Royal Commission on Labor in India (hereafter RCLI) in 1930 stumbled upon an old, and seemingly intractable, problem: that out of the 1,648,000 acres held in total by plantation agencies, only a little more than 26 percent (or around 430,000 acres) had actually been cropped.[3] More serious allegations were to be leveled against the industry by Assam's revenue minister seven decades later – that more than 3050 hectares of the region's prime agricultural land were under its "illegal" occupation.[4] When threatened with steep penalties, tea companies reportedly expressed helplessness in producing land records – for they were either unavailable or held up in London! While the political circumstances of this dispute in 2003 is beyond our purview, the point here is that the Indian takeover of plantation operations after 1947 did little to remedy these colonial legacies.[5]

This book also shows that the tea enterprise in Assam had never been amenable to standardized procedures or lawful conduct. And this perpetual "state of exception" was both institutionally sanctioned, and self-imposed. If the colonial government pandered to its extraordinary requests in the form of penal legislations and revenue-free lands, the industry was unhesitant to take matters in its own hands when it came to wage manipulation, worker welfare, or the more visceral aspects of labor control. Whereas direct forms of planter violence and regimes of authority are easy to detect (and have been repeatedly commented upon in existing scholarship), this work draws attention to the less visible, and historically ignored forms and legacies of these disorderly Edens. To this last group belongs the plantation's role in creating pathogen ecologies in the province – for humans and the tea plant; the industry's expedient use and abuse of categories such as "industrial" and "agrarian" in deciding labor duties or evading taxes; and the impunity with which law was

---

[2] On these issues, see Amalendu Guha, *Planter Raj to Swaraj: Freedom Struggle and Electoral Politics in Assam, 1826–1947* (New Delhi: ICHR, 1977), Guha, *Medieval and Early Colonial Assam* (Calcutta: K P Bagchi & Co., 1991), Arupjyoti Saikia, "State, Peasants and Land Reclamation: The Predicament of Forest Conservation in Assam, 1850s–1980s," *Indian Economic and Social History Review* 45, 1 (2008): 77–114, Saikia, "Forest Land and Peasant Struggles in Assam, 2002–2007," *Journal of Peasant Studies* 35.1 (2008): 39–59, and Debarshi Das and Arupjyoti Saikia, "Early Twentieth Century Agrarian Assam: A Brief and Preliminary Overview," *Economic and Political Weekly*, Vol. 46, No. 41 (October 8–14, 2011): 73–80.
[3] Ibid., p. 350.
[4] See Udayon Misra, "Assam Tea: The Bitter Brew," *Economic and Political Weekly*, Vol. 38, No. 29 (July 19–25, 2003): 3029–3032.
[5] For an ethnographic treatment of some of these questions, see Piya Chatterjee, *A Time for Tea: Women, Labor, and Post/Colonial Politics on an Indian Plantation* (Durham, NC and London: Duke University Press, 2001).

Conclusion 197

trampled and trespassed in matters ranging from labor welfare to forest felling. That the peculiarities of tea ecology lent itself to, or was manipulated to suit, plantation interests – especially in terms of worker compensation, or helped in the making of a malarial and bug-infested space – is part of the problem of non-regulation explored in this book.

The industry's self-characterization as an "agricultural" rather than an "industrial" undertaking had ramifications much beyond the tax issue discussed in Chapter 2. As far as these plantations are concerned, the RCLI's investigations revealed significant legal looseness and operational abuse of these categories. While Assam was repeatedly classified as an agricultural rather than an industrial province in this report, 549 of the province's 591 registered factories in 1929 were located within its tea estates.[6] Despite this, the manufacturing units with tea gardens were regularly kept outside the purview of the Indian Factories Act (IFA). For instance, the regulation of providing latrines in factories under the IFA was relaxed in the case of tea, as was the requirement of providing ventilation and fans in areas of high industrial dust. Similarly, while children were prohibited from employment in industrial tasks under the IFA, workers in the 7–9 age group were regularly used in tea-processing under the pretext of agricultural work.[7] When it came to the question of regularizing wage disbursement mechanisms in the tea industry – then ad hoc, variable from one estate to the next, and comprised of piece-work and task-work, as discussed – the Indian Tea Association and the planter's representative invoked these categories once again. They deposed before the RCLI that wage-fixing was unheard of and exceptional for "agricultural labor," and that such measures were more suitable for "sweated industries" where organization was faulty.[8]

The tea enterprise's unabashed claims, and confidence in its own state of order rarely held up to close scrutiny. Numerous extra-market, extra-legal, and even extra-human tactics and maneuvers belied these declarations. While labor historiography on the Assam plantations is replete with details on the brutal structures of power that kept workers and profits in place, persistent legal oversights – by the tea establishment and the government – were not inconsequential in this regard. Indeed, as we have seen throughout, the colonial administration intervened and looked askance at this issue depending on circumstance. If law, or lawlessness as I have argued, formed a distinctive component of this aforementioned state of exception, there were moments of scathing disclosure. A witness

---

[6] Vide *Royal Commission on Labour in India*, Evidence, Vol. VI [Assam and the Dooars] (London: His Majesty's Stationery Office, 1931), p. 8.
[7] Ibid., p. 205.   [8] Cited in *Report of the Royal Commission on Labour in India*, p. 403.

representing tea-recruiting interests testified to the RCLI that several provisions of the dreaded Act VI of 1901 were of doubtful validity, and that it needed "only once to be taken to the High Court to have many of its hollowness instantly exposed."[9] Despite these admissions, and recommendations for the system's overhaul by the labor inquiry commissions, legal elasticity remained a *sine qua non* of the Assam plantations. The withering portrayal of this notorious planter country in the vernacular press and literature along with periodic debates on the "return of slavery in the British dominion" in the House of Commons seldom changed anything. The lack of statistical standardization, accuracy, or completeness on many aspects of individual estate operations – mortality figures, acreage, worker earnings, methods of wage disbursement, worker demography, crop age – meant that a *uniform* picture of this economy in Assam was near impossible. Amidst such variability, law was a convenient alibi to either bolster company profits (by using strict penal clauses, for instance) or direct blame (with regard to worker sanitary regulations, for example). The remarkable aspect of these plantations was that it was a truly divided house, a free-for-all system, and to each his own – whether planter, agency house, or worker. While the semblance of industry unity, or default subscription to an overarching entity (the ITA or its Assam units) emerged during moments of crisis, annual meetings, and fiscal emergency, such privileges of union were denied to workers till the years before India's independence. To be sure, the first semblance of trade unions representing labor interests only emerged in the province around 1943.[10]

This book has also drawn attention to ideology's role in this plantation economy. This was by no means a benign abstraction, nor was its self-avowed mandate limited to botanical "improvement" in Assam. Throughout the period under study, and into the postcolonial period, an elaborate discourse of intellectual infancy legitimized labor's supposed inability to represent themselves.[11] The ventriloquism of opinion – by planters, the ITA, administrators, sanitarians, and later by union

---

[9] Ibid., p. 364.
[10] See Guha, *Planter Raj to Swaraj*, and Rana Partap Behal, *One Hundred Years of Servitude: Political Economy of Tea Plantations in Colonial Assam* (New Delhi: Tulika Books, 2014) for a discussion on trade union activity in the Assam gardens.
[11] Consider that when it came to the issue of education or hygiene, planters and sanitarians regularly invoked the idea of the laboring class as a primitive race, content only with subsistence work and a low standard of living. Even as late as 2003, primary education amongst tea labor children remained at an abysmal rate of 10 percent. More than 15 percent of tea estates had no lower primary schooling facility; cited in Misra, "Assam Tea: The Bitter Brew," pp. 3030–3031. The RCLI raised an interesting paradox in this regard: that planters regularly complained about innate worker "apathy and hostility" towards their own education, but invoked the excuse of intellectual infancy to

representatives and provincial governments – regularly drowned out workers' own view of life and times in these plantations. As a witness said without compunction to the RCLI: "the evolution of the coolie as a citizen [is only possible] by Government compulsion."[12] In dealing with sources on labor health, hygiene, compensation, welfare and overall well-being, therefore, I have been acutely aware that an unmediated access to these conditions, especially in the colonial period, is neither possible nor historically accurate. This book's reliance on colonial and so-called "elite" sources, and the sparseness of vernacular ones, testify to this reality. The task here has been to highlight the many elisions, the many silences, the many contradictions, and the many irregularities that fall within the two extreme pictures on labor in these plantations: that they were either in a state of perfect managerial bliss or subject to constant planter tyranny on the one hand, or "incapable and apathetic to change" on the other. For neither position is helpful to understand what went into the making of labor impoverishment and control in Assam, how they were put in place, and what was manipulated to ensure its success. For the tea enterprise, these strategies involved the political, the agronomic, the epidemiological, and the legal.

The doctrine of "improvement," of course, turned out to be a spectacular oxymoron in eastern India. Epidemiologically, the tea economy left devastating legacies in the region. Disease ecologies of malaria, black-fever, and cholera in these gardens stemmed from a combination of legal oversights, environmental disturbance, and reckless profiteering. Dr. G. C. Ramsay, Medical Officer of the Labac Medical Practice testified to the RCLI on January 4, 1930 that "most of the malaria in Assam has been unwittingly created by mankind."[13] Ramsay was categorical that the unchecked take-over of forest lands by tea speculators, large-scale tree felling, and injudicious irrigation and embankment works by the enterprise had exponentially increased *Anopheles minimus* in the province. For the tea plant, the micro-climate needed for its successful growth invited legions of pests and bugs that, in turn, led to reduced yields and loss of flavor. If tea bugs have remained an operational headache for planters and the management into the present, methods of control and eradication led to newer problems – not least in the use of harmful pesticides and herbicides.

---

deny them access to trade unionism; see *Royal Commission on Labour in India*, Evidence, Vol. VI [Assam and the Dooars], pp. 182–183.

[12] See *Royal Commission on Labour in India*, Evidence, Vol. VI [Assam and the Dooars], p. 183.

[13] Cited in *Royal Commission on Labour in India*, Evidence, Vol. VI, Part II, p. 10.

As it is, the tea economy had very little perceptible impact on Assam's agronomic growth in the colonial period. The situation did not change much with the transfer of ownership to Indian hands in 1947. Indeed, the Fifth Report of the Employment Review Committee, 1976 set up by the Assam Assembly noted wryly that almost all managerial posts of the 620 gardens in the Brahmaputra valley was held by "persons from outside the state [...] without any open advertisement or without notifying the employment exchange."[14] The province was also reported to have received a miniscule portion of the agricultural tax on 60 percent of tea profits in 1980, and only 22 crore of sales tax on tea sold from her 756 gardens during the same period.[15] The neighboring state of West Bengal was remarked to have netted close to 42 crore of these taxes through tea auction centers in Calcutta.[16] With almost all major tea companies headquartered in the former imperial capital, Assam was categorized as a "colonial hinterland" in these post-independent commercial games. It was this angst, among others, of being economically slighted and drained by a neo-colonial power structure based in New Delhi that gave rise to Assam's sustained right to self-determination movement in the late 1970s and the birth of secessionist militancy thereafter.[17] The long shadow of tea's purported "progressive" agenda was grim to say the least.

What of the Tea Bugs?

As actors go, insects, microbes and weeds make for less-than-perfect heroes, but they deserve their place in our narratives of the global plantation – and not just because you probably care about that daily glass of orange juice.[18]

---

[14] Cited in Tilottoma Misra, "Assam: A Colonial Hinterland," *Economic and Political Weekly*, Vol. 15, No. 32 (August 9, 1980): 1357–1364.
[15] In the Indian numbering system, a crore is equal to 10,000,000 or ten million.
[16] Ibid., pp. 1359–1361.
[17] On this, see Jayeeta Sharma's recent study, *Empire's Garden: Assam and the Making of India* (Durham, NC and London: Duke University Press, 2011), especially Part II; also, Sharit K. Bhowmik, "Ethnicity and Isolation: Marginalization of Tea Plantation Workers," *Race/Ethnicity: Multidisciplinary Global Contexts*, 4.2 (Winter 2011): 235–253, Udayon Misra, "Adivasi Struggle in Assam," *Economic and Political Weekly*, Vol. 42, No. 51 (December 22–28, 2007): 11–14, and Udayon Misra, *The Periphery Strikes Back: Challenges to the Nation-State in Assam and Nagaland* (Shimla: Indian Institute of Advanced Study, 2000), and Myron Weiner, "The Political Demography of Assam's Anti-Immigrant Movement." *Population and Development Review*, 9.2 (1983): 279–292.
[18] Frank Uekötter, ed., *Comparing Apples, Oranges, and Cotton: Environmental Histories of the Global Plantation* (Frankfurt, New York, NY: Campus Verlag, 2014), p. 25.

Managing plantation operations and pest bionomics were two sides of the same coin. Despite early warnings, the all-powerful ITA only took serious stock of the situation by the end of 1899, by appointing Harold H. Mann as its Chief Scientific Officer. By 1904, a full-fledged experimental station had come up in Heeleeka in upper Assam to assess cultivation methods, manuring techniques, provide entomological training to planters, and to devise methods to ensure crop healthiness. In an address at the 1911 annual meeting, the chairman of the ITA argued that a significant increase in total tea yields (from 400 lb in 1903 to 487 lb by the end of 1909) had to do with improved methods of working popularized by its Scientific Officers.[19] But Heeleeka had begun to prove inadequate, not least because of the paucity of "expertise" and the scattered location of officers all over the province. By 1911 a centralized center at Tocklai, also in upper Assam, had come up that consolidated the knowhow of entomologists, mycologists and experiments on tea "science" under one roof. At the time, Tocklai was made up of two bungalows, one laboratory and a sprawling hundred acres for crop testing. Chemical analysis, soil assessments, and fertilizer action against tea pests had all but commenced in Tocklai when the First World War curtailed its scientific ambitions.

The interwar period once again saw fervent technical activity in this regard. In 1922, a new entomological laboratory was set up in Tocklai followed by mycological and bacteriological ones the next year. Interestingly, the gap between esoteric knowledge and field expertise remained a recurring complaint during much of this period. Efforts to bridge the divide only unearthed more problems than solve any. For instance, in 1923, doubts arose about the efficacy of deep hoeing, then wildly popular among planters. The following year, tea men were advised against using lime as fertilizers despite contrary recommendations around 1910. Amidst these, the *Helopeltis theivora* or the tea mosquito bug continued to ravage techno-scientific confidence in pest control measures. In 1930, for instance, W. Y. Wyndham, planter in the Terai region complained bitterly that not enough had been done to thoroughly study the bug. Wyndham further asserted that he had incurred losses in the range of 10 to 30 percent per annum because of the pest.[20]

These contradictory findings, and individual protestations against the value of laboratory-based research, were not inconsequential. For the F. L. Engledow Commission, instituted to inquire into the workings of the ITA's scientific department, came to a similar conclusion. In its

---

[19] Cited in Sir Percival Griffiths, *The History of the Indian Tea Industry* (London: Weidenfeld and Nicolson, 1967), p. 436.
[20] Ibid., p. 443.

Report of March 3, 1936, the Commission noted that there had been insufficient contact between scientific prescription and on-site application. The tendency to apply general entomological results without regard to regional and intra-plantation variations was also criticized, as was the lack of scientific personnel in the many tea districts of Assam.[21] In other words, the practical considerations of entomological science and the pragmatic contingencies of tea-making had not necessarily come together all these years. The Engledow Commission was categorical that techno-scientific lines of inquiry needed to focus more on "factors affecting [tea] quality."[22]

Readers will note that despite these institutional advances, Samuel Peal's early caution discussed in Chapter 3 was remarkably prescient. For not only were planters, and the scientific community, unable to rid the industry of tea pests, the "single species forests" that plantations resembled provided an ever-attractive microclimate for their growth. In fact, out of the 1000 and more species of arthropods that infest tea globally[23], more than 380 come from India alone.[24] Even as late as 2002, studies revealed that crop loss due to these pests varied between 15 and 20 percent.[25] Routine use of chemical pesticides, averaging around 11.5 kilograms per hectare in the Assam gardens, for instance, came with attendant problems.[26] Among these were the resurgence of primary pests, secondary pest outbreaks, impedance in natural regulation, lethal and non-lethal impact on non-target organisms (including humans), undesirable residues in final product, and resistance to chemicals. For instance, a scientific finding in 2008 confirmed that *Helopeltis* was practically immune to synthetic pyrethroids, organophosphates, neonicotinoids, and organochlorine.[27] The red spider mite, originally discovered in Assam around 1868,[28] is now spread over multiple tea growing regions that include Bangladesh, Sri Lanka, Kenya, Taiwan, and Zimbabwe.[29]

---

[21] See Griffiths, *The History of the Indian Tea Industry*, pp. 445–450.   [22] Ibid., p. 448.
[23] Cited in L. K. Hazarika, M. Bhuyan and B. N. Hazarika, "Insect Pests of Tea and Their Management," *Annual Revenue of Entomology* 54 (2009): 267–284.
[24] See S. Roy, N. Muraleedharan, A. Mukhopadhyay, "The Red Spider Mite, *Oligonychus Coffeae* (Acari: *Tetranychidae*): its Status, Biology, Ecology and Management in Tea Plantations," *Exp. Appl. Acarol.* 63.4 (2014): 431–463.
[25] Quoted in N. Muraleedharan and R. Selvasundaram, "An IPM Package for Tea in India," *Planters Chronicle* 98 (2002): 107–124.
[26] These figures are for 1996; cited in G. Gurusubramanian et al., "Pesticide Usage Pattern in Tea Ecosystem, Their Retrospects and Alternative Measures," *Journal of Environmental Biology* 29.6 (2008): 813–826.
[27] Ibid., p. 815.
[28] See Sir George Watt's *The Pests and Blights of the Tea Plant* (Calcutta, 1898).
[29] Quoted in L. K. Hazarika, M. Bhuyan and B. N. Hazarika, "Insect Pests of Tea and their Management," p. 270.

It is interesting to note that though these recent findings are based on studies by entomologists and tea scientists working largely in Indian institutions (primarily at Tocklai but also elsewhere), their epistemic and practical concerns link them squarely to their colonial predecessors like Peal, Wood-Mason, and George Watt. While this bug question might now be a matter of India's national worry, their roots go back much into the past. Categories such as empire and colony, colonial and postcolonial are therefore fluid and transactional in this commodity story. It is in this sense that this book calls for the sociology of "ruination" to understand imperial projects and processes that continue to reside in, and inform, our contemporary histories.[30]

The pest issue, of course, was only part of a multi-faceted problem that included climate and rainfall. Just a decade after India's independence, experts concluded that plant disease was neatly correlated with crop age. The relationship was peculiar, for loss due to pests in the middle-age group (15–25) was about half of that in its early or mature stages. Figures for 1959, for instance, indicate that tea plants of 35 years and above in the Assam valley lost about 5–6 percent of their overall yield due to bugs and pests alone.[31] But bio-control mechanisms had rarely advanced since Peal's time. Integrated Pest Management (IPM) systems continue to advocate a mix of structural and synthetic approach: adjusting pruning cycles to occur every two to six years, intensive plucking, bush hygiene and weed removal, soil fertility treatment, trap crops, effective drainage, hand destruction alongside a wide array of biochemical fertilizers and botanicals.[32] Along with physical exposure to harmful bio-chemicals – by all measure an environmental justice issue – these systems of pest management often led to increased workloads without appreciable wage raise for workers in these plantations.[33] In recent times, stringent quality controls on pesticide use in tea by regulatory bodies such as the Food and

---

[30] See Ann L. Stoler's recent work on this concept in *Duress: Imperial Durabilities in Our Times* (Durham, NC: Duke University Press, 2016), and Stoler, ed., *Imperial Debris: On Ruins and Ruination* (Durham, NC: Duke University Press, 2013).

[31] See A. R. Sen and R. P. Chakrabarty, "Estimation of Loss of Crop from Pests and Diseases of Tea from Sample Surveys," *Biometrics* 20.3 (September 1964): 492–504.

[32] See L. K. Hazarika, M. Bhuyan, B. N. Hazarika, "Insect Pests of Tea and Their Management," and Somnath Roy et al. "Use of Plant Extracts for Tea Pest Management in India," *Appl. Microbiol. Biotechnol.* 100 (2016): 4831–4844, and G. Gurusubramanian et al., "Pesticide Usage Pattern in Tea Ecosystem, Their Retrospects and Alternative Measures," for an assessment.

[33] For Darjeeling, see Sarah Besky's recent work *The Darjeeling Distinction: Labor and Justice on Fair-Trade Tea Plantations in India* (Berkeley, CA and London: University of Berkeley Press, 2014); on the question of social and environmental justice, see Jill Harrison, *Pesticide Drift and the Pursuit of Environmental Justice* (Cambridge, MA: MIT Press, 2011).

Agricultural Organization (FAO), the Environmental Protection Agency (EPA), the World Health Organization (WHO), German Law (GL) and the European Economic Commission (EEC) and Indian governmental agencies has left tea companies scrambling for botanical alternatives. These regulatory standards, measured in terms of Minimum Residue Levels (MRL) in finished tea dictate export volume, price, and plantation credibility in the international market.[34] The socio-economic impact of these pest control measures on the end-producer – or tea laborer in our case – is a little-known story. Indeed, the social history of entomological "science" and its long-term effects on worker welfare is still to be written. To use Ann Stoler's words, "the temporal stretch, the uneven sedimentation of expanded and contracted forms of colonial effects, the grossly disparate ways in which that presence adheres to some bodies ... constitute one horizon of work to be done."[35]

Though tea produced from India regularly remains within prescribed MRLs, studies indicate that the subcontinent's problem with pesticides is more acute than others. Non-prescribed use, inappropriate supply to planters, sub-standard chemicals, lack of pedagogic training, and awareness among plantation functionaries have all been cited as reasons for this state of affairs.[36] In any case, phytochemicals remain a desirable, though little-used, alternative, comprising no more than 1 percent of the global insecticide market according to a recent finding.[37] However, the cost of organic solvents, lack of wider options, and lack of field-based, and species-wide effectiveness constrain their wholesale adoption in the Indian tea industry. As with Wood-Mason's experiments earlier, the use of botanical insecticides against tea pests (except *Azadirachta indica*, or neem) has remained at the level of laboratory research, or of academic interest at best.

This book invites readers to ponder over these facets of the Assam plantations – a decidedly transnational economic institution of the nineteenth and twentieth centuries. But while the plantation *form* made for its transferability across the Old World and the New, it was only through its constellation of local practices – and legacies – that these cash-crop economies impacted historically specific regions and communities. For if commodity capitalism and exchange provide a "unified conceptual

---

[34] See G. Gurusubramanian et al. "Pesticide Usage Pattern in Tea Ecosystem, Their Retrospects and Alternative Measures," for MRL figures in tea, and for a wider discussion of its impact on pesticide use in India.
[35] Stoler, *Duress*, p. 67.   [36] Ibid., p. 817.
[37] Vide Somnath Roy et al. "Use of Plant Extracts for Tea Pest Management in India," p. 4832.

field,"[38] it is only through the congeries of praxis, actors, and logics that we see its distinctive historical face – an *Assam* alongside a Darjeeling, Kenya, Sri Lanka, Java, Mauritius, Fiji, or Demerara. This book is in agreement with Timothy Mitchell's argument that the *idea* of a market "economy" cannot be understood as a self-contained and immanent category, an a priori objective unit of social analysis. Rather, as Mitchell suggests, economies were *products* of its internal dynamics, failures, and alliances – human and the nonhuman.[39] The disarray in Assam's imperial Edens, and its continuation into the present, points to these relationships, networks, and operational tactics in the making and unmaking of a truly global enterprise.

---

[38] For an argument espousing "conceptual universality" of market capitalism, see Andrew Sartori, *Bengal in Global Concept History: Culturalism in the Age of Capital* (Chicago, IL and London: The University of Chicago Press, 2008); also, Moishe Postone, *Time, Labor, and Social Domination: A Reinterpretation of Marx's Critical Theory* (Cambridge: Cambridge University Press, 1996).

[39] See Timothy Mitchell, *Rule of Experts: Egypt, Techno-Politics, Modernity* (Berkeley, CA: University of California Press, 2002).

Bibliography

Primary Sources

I. Archives

The Assam State Archives (ASA), Guwahati, Assam, India
Assam Secretariat Proceedings, home department
Assam Secretariat Proceedings, revenue department (A and B)
Assam Secretariat Proceedings, emigration department (A and B)
Assam Commissioner's Files – Land Revenue
Board of Revenue – Survey and Settlement Papers, Government of Eastern Bengal and Assam (1909–11)
Bengal Government Papers, Emigration
Evidence Recorded by the Forest Enquiry Committee, Assam for the Year 1929
Report on the Earthquake of the 12th June 1897, No. 5409G/A4282

Jorhat District Record Room, Office of the Collector and Deputy Commissioner, Jorhat, Assam, India
Deputy Commissioner's Files, Revenue department

Dibrugarh District Record Room, Office of the Collector and Deputy Commissioner, Dibrugarh, Assam, India
Deputy Commissioner's Files, Immigration department

Asian and African Studies (formerly India Office Records), British Library, London
The Indian Planters' Gazette and Sporting News (microfilm, MFM. MC1159)
European Manuscripts (Mss. Eur.)
    Government of India (GOI), Finance department

Indian Tea Association (ITA) circulars, reports, bulletins, and scientific manuals
A. J. W. Milroy Papers
Journals of Thomas Machell (1824–62)
Government of Eastern Bengal and Assam Bulletin – Agriculture department
Revenue Department Proceedings
Government of India-Home Miscellaneous (HM)
Public and Judicial Department Records (L/PJ)
Industries and Overseas Department Records (I&O)
Official Publications of the India Office (IOR/V)
India Office Economic Department Records (IOR/E)

National Library of Scotland Manuscript Collection, Edinburgh, Scotland
Diary of David Foulis, *The Tea Assistant in Cachar*, MS 9659

Center for South Asian Studies, Cambridge University, Cambridge
Lady B. Scott Papers, Box II

USDA National Agricultural Library, Beltsville, Maryland
The Indian Planters' Gazette and Sporting News (1915–22)

Center for Research Libraries (CRL), Chicago
The Amrita Bazar Patrika (1920–21)

II. Newspapers

The Gazette of India
Journal of the Society of Arts
Journal of the Agricultural and Horticultural Society of India
Calcutta Englishman
The Indian Forester
Assam District Gazetteer, Vol. 2 (Calcutta, 1905)
The Lancet
Medico-Chirurgical Transactions
Notes and Records of the Royal Society
Scientific Memoirs by Officers of the Medical and Sanitary Departments of the Government of India, New Series
Transactions of the Royal Society of Tropical Medicine and Hygiene

British Medical Journal
The Indian Medical Gazette
Pall Mall
The Northern Whig
The Bombay Chronicle

III. Published Primary Sources, Government Reports, Handbooks, and Tea Manuals (before c. 1930)

A Collection of the Acts of the Indian Legislature for the Year 1927 (Calcutta: Government of India Central Publication Branch, 1928).

Allen, B. C., *Assam District Gazetteer*, Vol. 2 (Calcutta, 1905).

Andrews, E. A., *Factors Affecting the Control of the Tea Mosquito Bug [Helopeltis theivora Waterhouse]* (London, n.d., Calcutta: ITA, rpt. 1910).

Annual Reports on the Administration of the Province of Assam (Calcutta: Secretariat Press).

Annual Sanitary Report of the Province of Assam for the Years 1876–82 (Shillong: Assam Secretariat Printing Office).

Baden-Powell, Baden H., *Forest Law: A Course of Lectures on the Principles of Civil and Criminal Law and on the Law of the Forest; Chiefly Based on the Laws in Force in British India* (London: Bradbury, Agnew & Co., 1893).

——— *The Land-Systems of British India: Being a Manual of the Land-Tenures and of the Systems of Land-Revenue Administration Prevalent in the Several Provinces*, Vol. III (Oxford: Clarendon Press, 1892).

——— "The Political Value of Forest Conservancy," *The Indian Forester*, Vol. II, No. 3, January 1877, 284.

Baildon, Samuel, *Tea in Assam: A Pamphlet on the Origin, Culture, and Manufacture of Tea in Assam* (Calcutta: W. Newman & Co., 1877).

Bald, Claud, *Indian Tea: Its Culture and Manufacture, Being a Textbook on the Culture and Manufacture of Tea*, Second edition (Calcutta: Thacker, Spink and Co., 1908).

Bamber, M. Kelway, *A Text Book on the Chemistry and Agriculture of Tea including the Growth and Manufacture* (Calcutta: Law Publishing Press, 1893).

Beadon Bryant, F., *A Note of Inspection on Some of the Forests of Assam* (Simla: Government Monotype Press, 1912).

Brandis, Dietrich, *Indian Forestry* (Woking: Oriental University Institute, 1897).

——— *Suggestions Regarding Forest Administration in the Province of Assam* (Calcutta: Superintendent of Government Printing, 1878).

Bruce, C. A., *An Account of the Manufacture of the Black Tea, As Now Practiced at Suddeya in Upper Assam* (Calcutta: Bengal Military Orphan Press, 1838).

Bulletin of Miscellaneous Information, (Royal Botanic Gardens, Kew, London: His Majesty's Stationery Office, 1906).

Carpenter, P. H. and C. J. Harrison, *The Manufacture of Tea in North-East India* (Calcutta: The Indian Tea Association, 1927).

Correspondence Regarding the Discovery of the Tea Plant of Assam, Proceedings of the Agricultural Society of India (Calcutta: Star Press, 1841).

Cotes, E. C., *An Account of the Insects and Mites Which Attack the Tea Plant in India* (Calcutta, 1895).
Cotton, J. H. S., *Indian and Home Memories* (London: T. Fisher Unwin, 1911).
Crawford, T. C., *Handbook of Castes and Tribes Employed on Tea Estates of North-East India* (Calcutta: Indian Tea Association, 1924).
Crole, David, *Tea: A Text Book of Tea Planting and Manufacture* (London: Crosby Lockwood and Son, 1897).
Day, Samuel Phillips, *Tea: Its Mystery and History* (London: Simpkin, Marshall & Co., 1877).
Deas, F. T. R., *The Young Tea Planter's Companion: A Practical Treatise on the Management of a Tea Garden in Assam* (London: S. Sonnenschein, Lowrey & Co., 1886).
Detailed Report of the General Committee of the Indian Tea Association for the Year 1880–1920 (Calcutta: Criterion Printing Works, 1921).
Dowling, A. F. (Compiler), *Tea Notes* (Calcutta: D. M. Traill, 1885).
Evidence Recorded by the Assam Labour Enquiry Committee (Shillong: Assam Government Press, 1922).
Evidence Recorded by the Forest Enquiry Committee, Assam for the Year 1929 (Shillong: Assam Government Press, 1929).
Fielder, C. H., "On the Rise, Progress, and Future Prospects of Tea Cultivation in British India," *Journal of the Statistical Society of London*, 32, 1 (March 1869), 29–37.
Giles, G. M., *A Report of an Investigation into the Causes of the Diseases Known in Assam as Kala-Azar and Beri-Beri* (Shillong: Assam Secretariat Press, 1890).
Grimley, W. H., *An Income Tax Manual Being Act II of 1886, with Notes* (Calcutta: Thacker, Spink & Co., 1886).
Haffkine, W. M., *Anti-Cholera Inoculation: Report to the Government of India* (Calcutta: Thacker, Spink & Co., 1895).
Protective Inoculation Against Cholera (Calcutta: Thacker, Spink & Co., 1913).
Hanley, Maurice, *Tales and Songs from an Assam Tea Garden* (Calcutta and Simla: Thacker, Spink and Co., 1928).
Hill, H. C., *Note on an Inspection of Certain Forests in Assam* (Calcutta: Office of the Superintendent of Government Printing, 1896).
Hope, G. D., *Memorandum on the Use of Artificial Manures on the Tea Estates of Assam and Bengal – Decade 1907–1917* (Calcutta: Star Printing Works, 1918).
Indian Law Reports, Calcutta Series, Vol. XLVIII, January to December (Calcutta: Bengal Secretariat Legislative Department, 1921).
M'Cosh, John, *Topography of Assam* (Calcutta: Bengal Military Orphan Press, 1837).
Mann, Harold H., *Early History of the Tea Industry of Northeast India* (Calcutta: General Printing Co. Ltd, 1918).
The Tea Soils of Cachar and Sylhet (Calcutta: The Indian Tea Association, 1903).
Marshall, Major G. F. L. and Lionel De Nicéville, *The Butterflies of India, Burmah and Ceylon* (Calcutta: The Calcutta Central Press, 1882).

Masters, J. W., "A Few Observations on Tea Culture," in *The Journal of the Agricultural and Horticultural Society of India*, Vol. III, Part I, January to December (Calcutta: Bishop's College Press, 1844).

Mills, A. J. Moffatt, Esq., *Report on the Province of Assam* (Calcutta: Gazette Office, 1854).

Money, Lieutenant-Colonel Edward, *The Tea Controversy (A Momentous Indian Question). Indian versus Chinese Teas. Which are Adulterated? Which are Better? With Many Facts about Both and the Secrets of the Trade* (London: W. B. Whittingham & Co., 1884).

The Cultivation and Manufacture of Tea, Third edition (London: W. B. Whittingham & Co., 1878).

Ovington, J., *An Essay upon the Nature and Qualities of Tea*, Second edition (London: Printed for John Chantry, 1705).

Papers Regarding the Tea Industry in Bengal (Calcutta: Bengal Secretariat Press, 1873).

Peal, S. E, "The Tea Bug of Assam," *Journal of the Agricultural and Horticultural Society of India* (New Series) 4(1) (1873), 126–132.

Price, J. Dodds and Leonard Rogers, "The Uniform Success of Segregation Measures in Eradicating Kala-Azar from Assam Tea Gardens: Its Bearing on the Probable Mode of Infection," *British Medical Journal*, 1(2771) (February 7, 1914), 285–289.

Progress Report of Forest Administration in the Province of Assam (Shillong: Secretariat Printing Office, 1876).

Report of the Assam Labour Enquiry Committee 1906 (Calcutta: Superintendent of Government Printing, 1907).

Report of the Assam Labour Enquiry Committee 1921–22 (Shillong: Government Press, 1922).

Report of the Commissioner of Patents for the Year 1860: Agriculture (Washington, DC: Government Printing Office, 1861).

Report of the Commissioner of Patents for the Year 1860: Agriculture, House of Representatives Papers, 36th Congress, 2nd Session, No. 48 (Washington, DC: Government Printing Office, 1861).

Report of the Commissioners Appointed to Enquire into the State and Prospects of Tea Cultivation in Assam, Cachar and Sylhet (Calcutta: Calcutta Central Press Company Ltd., 1868).

Report of the Indian Industrial Commission, 1916–1918 (London: His Majesty's Stationery Office, 1919).

Report on Labour Immigration into Assam for the Years 1876–1890 (Shillong: Assam Secretariat Press, 1891).

Report on Tea Culture in Assam for the Years 1878–1917 (Shillong: Government Press Assam, 1918).

Report on the Land Revenue Administration in the Province of Assam for the Year 1880 (Shillong: Secretariat Press, 1881).

Report on the Land Revenue Administration of the Lower Provinces for the Year 1870–71 (Calcutta: Government Press, 1872).

Rogers, Leonard, *Fevers in the Tropics, Their Clinical and Microscopical Differentiation, including the Milroy Lectures on Kala-Azar* (London: Oxford University Press, 1908).

"On the Epidemic Malarial Fever of Assam or Kala-Azar," *Medico-Chirurgical Transactions* 81, 1 (1898), 241–258.

Report of an Investigation of the Epidemic of Malarial Fever in Assam or Kala-Azar (Shillong: Assam Secretariat Printing Office, 1897).

Royal Commission on Agriculture in India: Evidence taken in Assam, Vol. V (London: His Majesty's Stationery Office, 1927).

Rules under the Inland Emigration Act I of 1882 (Calcutta: The Bengal Secretariat Press, 1884).

Schlich, W., *Manual of Forestry*, Vol. I (London: Bradbury, Agnew & Co., 1906).

Scientific Memoirs by Officers of the Medical and Sanitary Departments of the Government of India, New Series, No. 35 (Simla: Government Monotype Press, 1908).

Shipp, H. A., *Prize Essay on the Cultivation and Manufacture of Tea in Cachar* (Calcutta, 1865).

Sigmond, G. G., *Tea: Its Effects, Medicinal and Moral* (London: Longmans, 1839).

Strickland, C. and K. L. Chowdhury, *Abridged Report on Malaria in the Assam Tea Gardens: With Pictures, Tables and Charts* (Calcutta: Indian Tea Association, 1929).

The Assam Code: Containing the Bengal Regulations, Local Acts of the Governor General in Council, Regulations Made under the Government of India Act, 1870, and Acts of the Lieutenant-Governor of Bengal in Council, in Force in Assam, and Lists of the Enactments Which Have Been Notified for Scheduled Districts in Assam under the Scheduled Districts Act (Calcutta: Office of the Superintendent of Government Printing, 1897).

The Assam Forest Manual, Vol. I (Shillong: Government Press, 1923).

The Tea Cyclopaedia: Articles on Tea, Tea Science, Blights, Soils and Manures, Cultivation, Buildings, Manufacture Etc., With Tea Statistics (London: W. B. Whittingham & Co., 1882).

The Tea Planter's Vade Mecum (Calcutta: Office of the Tea Gazette, 1885).

The Unrepealed General Acts of the Governor General in Council, Vol. II [1864–1871] (Calcutta: Office of the Superintendent of Government Printing, 1876).

The Unrepealed General Acts of the Governor General in Council, Vol. III [1877–1881] (Calcutta: Office of the Superintendent of Government Printing, 1898).

The Unrepealed General Acts of the Governor General in Council: From Act I of 1914 to Act XI of 1919, Vol. VIII (Calcutta: Superintendent of Government Printing, 1919).

Thiselton Dyer, W. T., *The Botanical Enterprise of the Empire* (London: Eyre and Spottiswoode, 1880).

Transactions of the Royal Society of Tropical Medicine and Hygiene, 18, 3 (June 19, 1924), 81–86.

Watt, Sir George, *The Pests and Blights of the Tea Plant: Being a Report of Investigations Conducted in Assam and to Some Extent Also in Kangra* (Calcutta: Superintendent of Government Printing, 1898).

Wigley, F. G., *The Eastern Bengal and Assam Code: Containing the Regulations and Local Acts in Force in the Province of Eastern Bengal and Assam*, Vol. I (Calcutta: Superintendent of Government Printing, 1907).

Wood-Mason, James, *Report on the Tea-Mite and the Tea-Bug of Assam* (London: Taylor and Francis, 1884).

Secondary Sources

Adal, Kristin, "The Problematic Nature of Nature: The Post-Constructivist Challenge to Environmental History," *History and Theory*, Theme Issue 42 (December 2003), 60–74.

Anderson, Clare, *Subaltern Lives: Biographies of Colonialism in the Indian Ocean World, 1790–1920* (Cambridge: Cambridge University Press, 2012).

Antrobus, H. A., *A History of the Jorehaut Tea Company Ltd., 1859–1946* (London: Tea and Rubber Mail, 1947).

Arnold, David and Ramachandra Guha, eds., *Nature, Culture, Imperialism: Essays on the Environmental History of South Asia* (New Delhi: Oxford University Press, 1997).

Arnold, David, "Agriculture and 'Improvement' in Early Colonial India: A Pre-History of Development," *Journal of Agrarian Change*, Vol. 5, No. 4 (October 2005).

"British India and the 'Beri-Beri' Problem, 1798–1942," *Medical History* 54 (2010), 295–314.

"Plant Capitalism and Company Science: The Indian Career of Nathaniel Wallich," *Modern Asian Studies*, 42, 5 (2008), 899–928.

Colonizing the Body: State Medicine and Epidemic Disease in Nineteenth Century India (Berkeley, CA and London: University of California Press, 1993).

ed., *Warm Climates and Western Medicine: The Emergence of Tropical Medicine, 1500–1900* (Amsterdam: Rodopi B.V., 1996).

The Tropics and the Traveling Gaze: India, Landscape, and Science, 1800–1856 (Seattle, WA: The University of Washington Press, 2006).

Baer, Hans A., "On the Political Economy of Health," *Medical Anthropology Newsletter*, Vol. 14, No. 1 (November 1982): 1–2, 13–17.

Bambra, Clare, *Work, Worklessness, and the Political Economy of Health* (New York, NY: Oxford University Press, 2011).

Banerjee, Barundeb, "An Analysis of the Effects of Latitude, Age and Area on the Number of Arthropod Pest Species of Tea," *Journal of Applied Ecology* 18 (1981), 339–342.

Barpujari, H. K., *Assam: In the Days of the Company 1826–1858* (Gauhati: Lawyer's Book Stall, 1963).

The American Missionaries and North-East India, 1836–1900 (Guwahati: Spectrum, 1986).

The Comprehensive History of Assam, Vol. 4 (Gauhati: Assam Publication Board, 1992).

Barron, T. J., "Science and the Nineteenth-Century Ceylon Coffee Planters," *The Journal of Imperial and Commonwealth History* 16, 1 (1987), 5–23.

Barton, Gregory A., *Empire Forestry and the Origins of Environmentalism* (Cambridge: Cambridge University Press, 2007).
Baruah, Sanjib, *India against Itself: Assam and the Politics of Nationality* (New Delhi: Oxford University Press, 1999).
Baviskar, Amita, *In the Belly of the River: Tribal Conflicts over Development in the Narmada Valley*, Second edition (New Delhi: Oxford University Press, 2005).
Beckert, Sven, *Empire of Cotton: A Global History* (New York, NY: Vintage, 2014).
Behal, Rana Partap and Prabhu P. Mohapatra, "Tea and Money Versus Human Life: The Rise and Fall of the Indenture System in the Assam Tea Plantations 1840–1908," *Journal of Peasant Studies*, 19, 3 (1992), 142–172.
Behal, Rana Partap, *One Hundred Years of Servitude: Political Economy of Tea Plantations in Colonial Assam* (New Delhi: Tulika Books, 2014).
Beinart, William and Lotte Hughes, *Environment and Empire* (Oxford: Oxford University Press, 2007).
Berenbaum, May, *Bugs in the System: Insects and Their Impact on Human Affairs* (Reading, MA: Helix Books, 1995).
Besky, Sarah, *The Darjeeling Distinction: Labor and Justice on Fair-Trade Tea Plantations in India* (Berkeley, CA and London: University of Berkeley Press, 2014).
Bhadra, Gautam, Gyan Prakash and Susie Tharu, eds., *Subaltern Studies X: Writings on South Asian History and Society* (New Delhi: Oxford University Press, 1999).
Bhattacharya, Debraj, ed., *Of Matters Modern: The Experience of Modernity in Colonial and Post-colonial South Asia* (Calcutta: Seagull, 2008).
Bhattacharya, Nandini, "The Logic of Location: Malaria Research in Colonial India, Darjeeling, and Duars, 1900–30," *Medical History* 55 (2011), 183–202.
  Contagion and Enclaves: Tropical Medicine in Colonial India (Liverpool: Liverpool University Press, 2012).
Bhowmik, Sharit K., *Class Formation in the Plantation System* (New Delhi: People's Publishing House, 1981).
  'Ethnicity and Isolation: Marginalization of Tea Plantation Workers,' *Race/Ethnicity: Multidisciplinary Global Contexts*, 4.2 (Winter 2011), 235–253.
Bhuyan, B. and H. P. Sharma, 'Public Health Impact of Pesticide Use in the Tea Gardens of Lakhimpur District, Assam', *Ecology, Environment and Conservation* 10, 3 (2004): 333–338.
Bhuyan, S. K., *Early British Relations with Assam: A Study of the Original Sources and Records Elucidating the History of Assam from the Period of its First Contact with the Honourable East India Company to the Transfer of the Company's Territories to the Crown in 1858* (Shillong: Assam Govt. Press, 1949).
Bose, Sanat, *Capital and Labor in the Indian Tea Industry* (Bombay: All India Trade Union Congress, 1954).
Breen, T. H., *Tobacco Culture: The Mentality of the Great Tidewater Planters on the Eve of Revolution* (Princeton, NJ: Princeton University Press, 1985).

Broomfield, J. H., *Elite Conflict in a Plural Society: Twentieth Century Bengal* (Berkeley, CA: University of California Press, 1968).
Burton, Antoinette, ed., *Archive Stories: Facts, Fictions and the Writing of History* (Durham, NC: Duke University Press, 2005).
'Thinking beyond the Boundaries: Empire, Feminism and the Domains of History,' *Social History*, Vol. 26, No. 1 (2001), 60–71.
Chakrabarti, Pratik, 'Curing Cholera: Pathogens, Places and Poverty in South Asia,' *International Journal of South Asian Studies*, 3 (December 2010), 153–68.
Chatterjee, Indrani, *Forgotten Friends: Monks, Marriages, and Memoirs of Northeast India* (New Delhi: Oxford University Press, 2013).
Chatterjee, Piya, *A Time for Tea: Women, Labor, and Post/Colonial Politics on an Indian Plantation* (Durham, NC and London: Duke University Press, 2001).
Chatterjee, Rakhahari, "C. R. Das and the Chandpur Strikes of 1921," *Bengal Past and Present*, Vol. XCIII, Parts II & III, Serial No. 176 & 177 (September–December 1974), 181–196.
Chattopadhyay, Dakshinacharan, *Cha-kar Darpan Natak in Bangla Natya Sankalan* (Calcutta, 2001).
Chaturvedi, Benarsidas and Marjorie Sykes, *Charles Freer Andrews: A Narrative* (New Delhi: Govt. of India Publications Division, 1971).
Clark, J. F. M., *Bugs and the Victorians* (New Haven, CT and London: Yale University Press, 2009).
Cook, G. C., "Leonard Rogers KCSI FRCP FRS (1868–1962) and the Founding of the Calcutta School of Tropical Medicine," *Notes and Records of the Royal Society*, 60 (2006), 171–181.
Corrigan, Philip, "Feudal Relics or Capitalist Monuments? Notes on the Sociology of Unfree Labor," *Sociology* 11, 3 (1977), 435–463.
Cronon, William, "The Uses of Environmental History," *Environmental History Review*, 17, 3 (1993), 1–22.
Crosby, Alfred, *Ecological Imperialism: The Biological Expansion of Europe, 900–1900* (Cambridge: Cambridge University Press, 1986).
Das, Debarshi and Arupjyoti Saikia, "Early Twentieth Century Agrarian Assam: A Brief and Preliminary Overview," *Economic and Political Weekly*, Vol. 46, No. 41 (October 8–14, 2011), 73–80.
Das, G. M., "Bionomics of the Tea Red Spider, *Oligonychus Coffeae* (Nietner)," *Bulletin of Entomology*, 50, 2 (1959), 265–274.
Das Gupta, Ranajit, *Labor and Working Class in Eastern India: Studies in Colonial History* (Calcutta and New Delhi: K. P. Bagchi & Company, 1994).
Das, Rajani Kanta, *Plantation Labor in India* (Calcutta: Prabasi Press, 1931).
Drayton, Richard, *Nature's Government: Science, Imperial Britain, and the "Improvement" of the World* (New Haven: Yale University Press, 2000).
Dreitzel, Hans Peter, ed., *The Social Organization of Health*, Recent Sociology No. 3 (New York, NY and London: Macmillan, 1971).
Duncan, James S., *In the Shadows of the Tropics: Climate, Race and Biopower in Nineteenth Century Ceylon* (London: Ashgate Publishing Co., 2007).

Dunn, Richard S., *Sugar and Slaves: The Rise of the Planter Class in the English West Indies, 1624–1713* (Chapel Hill, NC: University of North Carolina Press, 1972).
Emmer, P. C., ed., *Colonialism and Migration; Indentured Labour before and after Slavery* (The Netherlands: Martinus Nijhoff, 1986).
Engels, Friedrich, *The Condition of the Working Class in England*, trans. W. O. Henderson and W. H. Chaloner (Stanford, CA: Stanford University Press, rpt. 1958).
Erikson, Emily, *Between Monopoly and Free Trade: The English East India Company, 1600–1757* (Princeton, NJ: Princeton University Press, 2016).
Foucault, Michel, *"Society Must be Defended": Lectures at the Collège de France, 1975–1976*, trans. David Macey (New York, NY: Picador, rpt. 2003).
The History of Sexuality, Vols. 1, 2 and 3, (New York, NY: Vintage).
Fraser, W. M., *The Recollections of a Tea Planter* (London: Tea and Rubber Mail, 1935).
Gadgil, Madhav and Ramachandra Guha, *The Use and Abuse of Nature* (New Delhi: Oxford University Press, 2005).
Gait, Edward, *A History of Assam* (Calcutta: Thacker, Spink, rpt. 1967).
Gandhi, Mohandas K., *The Collected Works of Mahatma Gandhi* (Delhi: Govt. of India Publications Division).
Ganguly, Dwarkanath, *Slavery in British Dominion*, Siris Kumar Kunda, ed. (Calcutta: Jijnasa Publications, 1972).
Gawthrop, W. R. (Compiler), *The Story of the Assam Railways and Trading Company Limited, 1881–1951* (London: Harley Pub. Co. for the Assam Railways and Trading Company, 1951).
Ghosh, Durba, "Another Set of Imperial Turns?" *American Historical Review*, Vol. 117, No. 3 (June 2012), 772–793.
Gilmartin, David, "Scientific Empire and Imperial Science: Colonialism and Irrigation Technology in the Indus Basin," *The Journal of Asian Studies*, 53. 4 (November 1994), 1127–1149.
Gohain, Hiren, "Politics of a Plantation Economy," Review of Amalendu Guha, *Planter's Raj to Swaraj* in *Economic and Political Weekly*, Vol. 13. No. 13 (April 1, 1978), 579–80.
Goswami, Shrutidev, "The Opium Evil in Nineteenth Century Assam," *Indian Economic and Social History Review*, Vol. XIX, No. 3 & 4 (1982), 365–376.
Griffiths, Percival, *The History of the Indian Tea Industry* (London: Weidenfeld and Nicolson, 1967).
Grove, Richard H., *Green Imperialism: Colonial Expansion, Tropical Island Edens and the Origins of Environmentalism, 1600–1860* (Cambridge: Cambridge University Press, 1995).
Guha, Amalendu, "Assamese Agrarian Relations in the Later Nineteenth Century: Roots, Structure and Trends," *The Indian Economic and Social History Review*, Vol. XVII. No. 1, January–March 1980.
"Imperialism of Opium in Assam 1773–1921," *Calcutta Historical Journal*, Vol. 1, No. 2 (January–June 1977), 226–245.

Planter Raj to Swaraj: Freedom Struggle and Electoral Politics in Assam 1826–1947 (New Delhi: ICHR, 1977).

Guha, Ramachandra, "Forestry in British and Post-British India: A Historical Analysis," *Indian Economic and Social History Review*, Part I & II, 18, 44 (October 29, 1983), 1882–1896.

"Forestry in British and Post-British India: A Historical Analysis,' *Indian Economic and Social History Review*, Part III & IV, 18, 45/46 (November 5–12, 1983).

The Unquiet Woods: Ecological Change and Peasant Resistance in the Himalaya (Ranikhet: Permanent Black, rpt. 2013).

Guha, Ranajit and Gayatri Chakravorty Spivak, eds., *Selected Subaltern Studies* (New Delhi: Oxford University Press, 1988).

Guha, Ranajit, *A Rule of Property for Bengal: An Essay on the Idea of Permanent Settlement* (Durham, NC and London: Duke University Press, rpt. 1996).

Gurusubramanian, G. et al. "Pesticide Usage Pattern in Tea Ecosystem, Their Retrospects and Alternative Measures," *Journal of Environmental Biology* 29.6 (2008), 813–826.

Guthman, Julie, *Agrarian Dreams: The Paradox of Organic Farming in California* (Berkeley, CA: University of California Press, 2004).

Habib, Irfan, *The Agrarian System of Mughal India 1556–1707*, Third edition (New Delhi: Oxford University Press, rpt. 2014).

Haijian, Mao, *The Qing Empire and the Opium War: The Fall of the Heavenly Dynasty*, trans. Joseph Lawson, Peter Lavelle and Craig Smith (Cambridge: Cambridge University Press, 2016).

Hall, Catherine and Sonya O. Rose, eds., *At Home with the Empire: Metropolitan Culture and the Imperial World* (Cambridge: Cambridge University Press, 2006).

Hall, Catherine, ed., *Cultures of Empire: Colonizers in Britain and the Empire in the Nineteenth and Twentieth Centuries—A Reader* (Manchester: Manchester University Press, 2000).

Haraway, Donna, *The Haraway Reader* (New York, NY: Routledge, 2004).

"The Promises of Monsters: A Regenerative Politics for Inappropriate/d Others," in Lawrence Grossberg, Cary Nelson and Paula Treichler, eds., *Cultural Studies* (New York, NY: Routledge, 1992), pp. 295–337.

Modest_Witness@Second_Millenium.FemaleMan©_Meets _OncoMouse™ (London: Routledge, 1997).

Harrison, Jill, *Pesticide Drift and the Pursuit of Environmental Justice* (Cambridge, MA: MIT Press, 2011).

Harrison, Mark, *Public Health in British India: Anglo-Indian Preventive Medicine, 1859–1914* (Cambridge: Cambridge University Press, 1994).

Hay, Douglas and Paul Craven, eds., *Masters, Servants, and Magistrates in Britain and the Empire, 1562–1955* (Chapel Hill, NC and London: The University of North Carolina Press, 2004).

Hazarika, L. K., M. Bhuyan and B. N. Hazarika, "Insect Pests of Tea and their Management," *Annual Review of Entomology* 54 (2009): 267–284.

Husserl, Edmund, *The Idea of Phenomenology*, trans. William Alston and George Nakhnikian (The Hague: Nijhoff, 1964).

Jaco, E. Gartly, ed. *Patients, Physicians, and Illness: A Sourcebook in Behavioral Science and Health*, Third edition (New York, NY: The Free Press, 1979).
Jha, J. C., *Aspects of Indentured Inland Emigration to North-East India 1859–1918* (New Delhi: Indus Publishing Company, 1996).
Kar, Bodhisatwa, *Framing Assam: Plantation Capital, Metropolitan Knowledge and a Regime of Identities, 1790s–1930s*, unpublished PhD dissertation (New Delhi: Jawaharlal Nehru University, 2007).
Keay, John, *The Honourable Company: A History of the English East India Company* (New York, NY: Macmillan, 1994).
Kelman, Sander, "Introduction to the Theme: The Political Economy of Health," *International Journal of Health Services*, 5, 4 (1975), 535–538.
Kelman, Sander, "The Social Nature of the Definition Problem in Health," *International Journal of Health Services*, 5, 4 (1975), 625–642.
Kirsch, Scott and Don Mitchell, "The Nature of Things: Dead Labor, Non-Human Actors, and the Persistence of Marxism," *Antipode* 36 (2002), 687–705.
Klein, Ira, "Development and Death: Reinterpreting Malaria, Economics, and Ecology in British India," *The Indian Economic and Social History Review* 38 (2001), 147–79.
Kohler, Robert E., *Lords of the Fly: Drosophila Genetics and the Experimental Life* (Chicago, IL: University of Chicago Press, 1994).
Kolsky, Elizabeth, *Colonial Justice in British India: White Violence and the Rule of Law* (Cambridge and New York, NY: Cambridge University Press, 2010).
Kumar, Deepak, *Science and the Raj, 1857–1905* (New Delhi: Oxford University Press, 1995).
Kumar, Dharma and Meghnad Desai, eds., *The Cambridge Economic History of India, Vol. 2, c. 1751–1970* (Cambridge: Cambridge University Press, 1983).
Kumar, Prakash, *Indigo Plantations and Science in Colonial India* (Cambridge and New York, NY: Cambridge University Press, 2012).
  "Plantation Indigo and Synthetic Indigo: European Planters and the Redefinition of a Colonial Commodity," *Comparative Studies in Society and History*, 58.2 (2016), 407–431.
Latour, Bruno, *Reassembling the Social: An Introduction to Actor-Network Theory* (New York, NY: Oxford University Press, 2007).
  We Have Never Been Modern, trans. Catherine Porter (Cambridge, MA: Harvard University Press, 1993).
Law, John and J. Hassard, eds., *Actor Network Theory and After* (Oxford: Blackwell, 1999).
Longley, P. R. H., *Tea Planter Sahib: The Life and Adventures of a Tea Planter in North East India* (Auckland: Tonson Publishing House, 1969).
Lutz, Catherine, "Empire Is in the Details," *American Ethnologist*, Vol. 33, No. 4 (November 2006), 593–611.
McIntyre, W. D., ed., *The Journal of Henry Sewell, vol. 1. February 1853 to May 1854* (Christchurch: Whitcoulls, 1980).
Majumdar, R. C., *History of the Freedom Movement in India*, Vol. III (Bombay: K L Mukhopadhyay, 1963).

Marx, Karl, *Capital*, Vol. I, trans. Ben Fowkes (New York, NY: Vintage, rpt. 1977).
Marx, Leo, *Machine in the Garden: Technology and the Pastoral Ideal in America* (New York, NY: Oxford University Press, 1964).
McNeill, J. R., *Mosquito Empires: Ecology and War in the Greater Caribbean, 1620–1914* (Cambridge: Cambridge University Press, 2010).
Merchant, Carolyn, *Autonomous Nature: Problems of Prediction and Control from Ancient Times to the Scientific Revolution* (New York, NY and London: Routledge, 2016).
  Reinventing Eden: The Fate of Nature in Western Culture (New York, NY and London: Routledge, 2003).
Miles, Robert, *Capitalism and Unfree Labor: Anomaly or Necessity?* (London: Tavistock Publications, 1987).
Mintz, Sidney W., *Sweetness and Power: The Place of Sugar in Modern History*, (New York, NY and London: Viking, 1985).
Misra, Tilottoma, "Assam: A Colonial Hinterland," *Economic and Political Weekly*, Vol. 15, No. 32 (August 9, 1980), 1357–1364.
Misra, Udayon, "Assam Tea: The Bitter Brew," *Economic and Political Weekly*, Vol. 38, No. 29 (July 19–25, 2003), 3029–3032.
  "Adivasi Struggle in Assam," *Economic and Political Weekly*, Vol. 42, No. 51 (December 22–28, 2007), 11–14.
  The Periphery Strikes Back: Challenges to the Nation-State in Assam and Nagaland (Shimla: Indian Institute of Advanced Study, 2000).
Mitchell, Timothy, *Rule of Experts: Egypt, Techno-Politics, Modernity* (Berkeley, CA and London: University of California Press, 2002).
Mitman, Gregg, *The State of Nature: Ecology, Community, and American Social Thought, 1900–1950* (Chicago. IL: University of Chicago Press, 1992).
Moxham, Roy, *Tea: Addiction, Exploitation, and Empire* (New York. NY: Carroll & Graf, 2003).
Mulcahy, Matthew, *Hurricanes and Society in the British Greater Caribbean, 1624–1783* (Baltimore, MD: The Johns Hopkins University Press, 2006).
Muraleedharan, N. and R. Selvasundaram, "An IPM Package for Tea in India," *Planters Chronicle* 98 (2002), 107–24.
Navarro, Vicente, ed., *Neoliberalism, Globalization and Inequalities: Consequences for Health and Quality of Life* (Amityville, NY: Baywood Publishers, 2007).
Navarro, Vicente, *Medicine under Capitalism* (New York, NY: Croom Helm Ltd., 1976).
Nelson, Lynn A., *Pharsalia: An Environmental Biography of a Southern Plantation, 1780–1880* (Athens, GA: University of Georgia Press, 2007).
Numbers, Ronald L. and Todd L. Savitt, eds., *Science and Medicine in the Old South* (Baton Rouge, LA and London: Louisiana State University Press, 1989).
Olmstead, Alan L. and Paul W. Rhode, *Creating Abundance: Biological Innovation and American Agricultural Development* (Cambridge and New York, NY: Cambridge University Press, 2008).

Pagar, S. M., *The Indian Income Tax: Its History, Theory, and Practice* (Baroda, 1920).
Pandhe, Pramila, ed., *Suppression of Drama in Nineteenth Century India* (Calcutta: India Book Exchange, 1978).
Parthesius, Robert, *Dutch Ships in Tropical Waters: The Development of the Dutch East India Company (VOC) shipping network in Asia, 1595–1660* (Amsterdam: Amsterdam University Press, 2010).
Pati, Biswamoy and Mark Harrison, eds., *The Social History of Health and Medicine in Colonial India* (Abingdon: Routledge, 2009).
Polu, Sandhya, *Infectious Disease in India, 1892–1940: Policy-Making and the Perception of Risk* (London: Palgrave Macmillan, 2012).
Postone, Moishe, *Time, Labor, and Social Domination: A Reinterpretation of Marx's Critical Theory* (Cambridge: Cambridge University Press, 1996).
Power, Helen, "The Calcutta School of Tropical Medicine: Institutionalizing Medical Research in the Periphery," *Medical History* 40 (1996), 197–214.
Prakash, Gyan, *Another Reason: Science and the Imagination of Modern India* (Princeton, NY: Princeton University Press, 1999).
Prest, John, *The Garden of Eden: The Botanic Garden and the Recreation of Paradise* (New Haven, CT and London: Yale University Press, 1981).
Raj, Kapil, *Relocating Modern Science: Circulation and the Construction of Knowledge in South Asia and Europe, 1650–1900* (Basingstoke: Palgrave, 2007).
Rajan, S. Ravi, *Modernizing Nature: Forestry and Imperial Eco-Development 1800–1950* (New Delhi: Orient Longman, rpt. 2008).
Rangarajan, Mahesh, *Fencing the Forest: Conservation and Ecological Change in India's Central Provinces 1860–1914* (New Delhi: Oxford University Press, 1996).
  "Environmental Histories of India: *Of States, Landscapes, and Ecologies*," in Edmund Burke III and Kenneth Pomeranz, eds., *The Environment and World History* (Berkeley, CA: University of California Press, 2009).
Rangarajan, Mahesh and K. Sivaramakrishnan, eds., *Shifting Ground: People, Mobility and Animals in India's Environmental Histories* (New Delhi: Oxford University Press, 2014).
Rappaport, Erika, *A Thirst for Empire: How Tea Shaped the Modern World* (Princeton, NJ and Oxford: Princeton University Press, 2017).
Richards, John F., *The Unending Frontier: An Environmental History of the Early Modern World* (Berkeley, CA and London: University of California Press, 2003).
Rogers, Leonard, "The Epidemic Malarial Fever of Assam, or Kala-Azar, Successfully Eradicated from Tea Garden Lines," September 24, 1898, *British Medical Journal*, 2 (1969): 891–892.
Rosenberg, Charles and Janet Golden, eds., *Framing Disease: Studies in Cultural History* (New Brunswick, NJ: Rutgers University Press, 1992).
Ross, Corey, *Ecology and Power in the Age of Empire: Europe and the Transformation of the Tropical World* (Oxford: Oxford University Press, 2017).
Rowe, William T., *China's Last Empire: The Great Qing* (Harvard, MA: Belknap Press of Harvard University Press, 2009).

Hankow: Commerce and Society in a Chinese City, 1796–1889 (Stanford, CA: Stanford University Press, 1992).

Rowntree, John, *A Chota Sahib: Memoirs of a Forest Officer* (Padstow: Tabb House, 1981).

Roy, S., N. Muraleedharan and A. Mukhopadhyay, "The Red Spider Mite, *Oligonychus Coffeae* (Acari: *Tetranychidae*): Its Status, Biology, Ecology and Management in Tea Plantations," *Exp. Appl. Acarol.* 63.4 (2014), 431–63.

Roy, Somnath, et. al. "Use of Plant Extracts for Tea Pest Management in India," *Appl. Microbiol. Biotechnol.* 100 (2016): 4831–4844.

Roy, Tirthankar, *The East India Company: The World's Most Powerful Corporation* (New Delhi: Allen Lane, 2012).

The Economic History of India, 1857–1947, Third edition (New Delhi: Oxford University Press, 2011).

Saikia, Arupjyoti, "Imperialism, Geology and Petroleum: History of Oil in Colonial Assam," *Economic and Political Weekly*, Vol. XLVI, 12 (March 19, 2011), 48–55.

A Century of Protests: Peasant Politics in Assam Since 1900 (New Delhi: Routledge, 2014).

"State, Peasants and Land Reclamation: The Predicament of Forest Conservation in Assam, 1850s–1980s," *Indian Economic and Social History Review* 45, 1 (2008), 77–114.

Forests and Ecological History of Assam, 1826–2000 (New Delhi: Oxford University Press, 2011).

ed., *Orunodoi: Collected Essays 1855–1868* [in Assamese] (Nagaon: Krantikaal Prakashan, 2002).

Saikia, Rajen, *Social and Economic History of Assam 1853–1921* (New Delhi: Manohar, 2000).

Samanta, Arabinda, *Malarial Fever in Colonial Bengal: Social History of an Epidemic, 1820–1939* (Kolkata: Firma KLM, 2002).

Sarkar, Tanika, *Rebels, Wives, Saints: Designing Selves and Nations in Colonial Times* (New Delhi: Permanent Black, 2009).

Sartori, Andrew, *Bengal in Global Concept History: Culturalism in the Age of Capital* (Chicago, IL and London: The University of Chicago Press, 2008).

Sartre, Jean-Paul, *Basic Writings*, Stephen Priest, ed. (London: Routledge, 2001).

Savitt, Todd L. and James Harvey Young, eds., *Disease and Distinctiveness in the American South* (Knoxville, TN: The University of Tennessee Press, 1988).

Schiebinger, Londa, *Plants and Empire: Colonial Bioprospecting in the Atlantic World* (Cambridge, MA and London: Harvard University Press, 2004).

Scott, James C., *Seeing Like a State: How Certain Schemes to Improve the Human Condition Have Failed* (New Haven, CT and London: Yale University Press, 1998).

The Moral Economy of the Peasant: Rebellion and Resistance in Southeast Asia (New Haven, CT: Yale University Press, 1976).

Sen, A. R. and R. P. Chakrabarty, "Estimation of Loss of Crop from Pests and Diseases of Tea from Sample Surveys," *Biometrics* 20.3 (September 1964), 492–504.

Sen, Samita, "Commercial Recruiting and Informal Intermediation: Debate over the Sardari System in Assam Tea Plantations, 1860–1900," *Modern Asian Studies* 44, 1 (2010), 3–28.

Shah, Alpa, *In the Shadows of the State: Indigenous Politics, Environmentalism, and Insurgency in Jharkhand, India* (Durham, NC: Duke University Press, 2010).

Sharma, Jayeeta, *Empire's Garden: Assam and the Making of India* (Durham, NC and London: Duke University Press, 2011).

Shiva, Vandana, *Staying Alive: Women, Ecology, and Development* (New Delhi: Kali for Women, 1988).

Shlomowitz, Ralph and Lance Brennan, "Mortality and Migrant Labour in Assam, 1865–1921," *The Indian Economic and Social History Review* 27, 1 (1990), 85–110.

Siddique, Muhammed Abu B., *Evolution of Land Grants and Labour Policy of Government: The Growth of the Tea Industry in Assam 1834–1940* (New Delhi: South Asian Publishers, 1990).

Singer, Merrill, "Developing a Critical Perspective in Medical Anthropology," *Medical Anthropology Quarterly*, 17, 5 (1986), 128–129.

Singha, Radhika, *A Despotism of Law: Crime and Justice in Early Colonial India* (New Delhi: Oxford University Press, 1998).

Sinha, Mrinalini, "Historia Nervosa or Who's Afraid of Colonial-Discourse Analysis," *Journal of Victorian Culture*, Vol. 2, 1 (1997), 113–122.

Sircar, Kalyan K., "A Tale of Two Boards: Some Early Management Problems of Assam Company Limited, 1839–1864," *Economic and Political Weekly* 21, 10/11 (March 1986), 453–459.

"Labor and Management: First Twenty Years of Assam Company Limited (1839–59)," *Economic and Political Weekly* 21, 22, (May 1986), M38–M43.

"Coolie Exodus from Assam's Chargola Valley, 1921: An Analytical Study," *Economic and Political Weekly*, Vol. 22, No. 5 (January 31, 1987), 184–193.

Sivaramakrishnan, K., *Modern Forests: Statemaking and Environmental Change in Colonial Eastern India* (Stanford, CA: Stanford University Press, 1999).

Soluri, John, *Banana Cultures: Agriculture, Consumption, and Environmental Change in Honduras and the United States* (Austin, TX: University of Texas Press, 2005).

Spivak, Gayatri Chakravorty, "The Rani of Sirmur: An Essay in Reading the Archives," *History and Theory* 24, No. 3 (1985), 247–272.

"Can the Subaltern Speak?" in Cary Nelson and Lawrence Grossberg, eds., *Marxism and the Interpretation of Culture* (Urbana, IL: University of Illinois Press, 1988), 271–313.

Sramek, Joseph, "'Face Him Like a Briton': Tiger Hunting, Imperialism, and British Masculinity in Colonial India, 1800–1875," *Victorian Studies*, 48, 4 (Summer 2006), 659–680.

Steedman, Carolyn, *Master and Servant: Love and Labour in the English Industrial Age* (Cambridge: Cambridge University Press, 2007).

Stoler, Ann Laura, *Capitalism and Confrontation in Sumatra's Plantation Belt, 1870–1979* (New Haven, CT and London: Yale University Press, 1985).

Duress: Imperial Durabilities in Our Times (Durham, NC: Duke University Press, 2016).
   ed., *Imperial Debris: On Ruins and Ruination* (Durham, NC: Duke University Press, 2013).
Sundar, Nandini, *Subalterns and Sovereigns: An Anthropological History of Bastar (1854–2006)*, Second edition (New Delhi: Oxford University Press, 2008).
Sutter, Paul S., "Nature's Agents or Agents of Empire? Entomological Workers and Environmental Change during the construction of the Panama Canal," *Isis* 98, 4 (2007), 724–754.
Thompson, E. P., *The Making of the English Working Class* (New York, NY: Vintage, rpt. 1966).
Thompson, Edgar T., *The Plantation: A Bibliography*, Social Science Monographs IV (Washington, DC: Pan American Union, 1957).
Tilley, Helen, *Africa as a Living Laboratory: Empire, Development, and the Problem of Scientific Knowledge, 1870–1950* (Chicago, IL: University of Chicago Press, 2011).
Tinker, Hugh, *A New System of Slavery: The Export of Indian Labor Overseas, 1830–1920* (London: Institute of Race Relations, 1974).
Tyrrell, Ian, *True Garden of the Gods: Californian-Australian Environmental Reform, 1860–1930* (Berkeley, CA: University of California Press, 1999).
Uekötter, Frank, ed., *Comparing Apples, Oranges, and Cotton: Environmental Histories of the Global Plantation* (Frankfurt and New York, NY: Campus Verlag, 2014).
Ukers, William H, *All about Tea*, Vol. I (New York, NY: The Tea and Coffee Trade Journal Company, 1935).
Varma, Nitin, "Coolie Acts and the Acting Coolies: Coolie, Planter and State in the Late Nineteenth and Early Twentieth Century Colonial Tea Plantations of Assam," *Social Scientist* Vol. 33, No. 5/6 (May–June 2005), 49–72.
   Producing Tea Coolies? Work, Life and Protest in the Colonial Tea Plantations of Assam, 1830s–1920s, unpublished D.Phil. dissertation (Berlin: Humboldt University, 2011).
Vidyaratna, Ramkumar, *Kuli Kahini*, Biswanath Mukhopadhyay, ed. (Calcutta: Jogomaya Publications, 1886).
Waitzkin, Howard, "The Social Origins of Illness: A Neglected History," *International Journal of Health Services*, 11, 1 (1981), 77–105.
   Medicine and Public Health at the End of Empire (Boulder, CO: Paradigm Publishers, 2011).
Waley, Arthur, *The Opium War through Chinese Eyes* (Stanford, CA: Stanford University Press, 1958).
Webb, James L. A. Jr., *Tropical Pioneers: Human Agency and Ecological Change in the Highlands of Sri Lanka, 1800–1900* (New Delhi: Oxford University Press, 2002).
Weiner, Myron, "The Political Demography of Assam's Anti-Immigrant Movement." *Population and Development Review*, 9.2 (1983), 279–92.
White, Richard, "Discovering Nature in North America," *Journal of American History* 79 (1992), 874–891.

The Organic Machine: The Remaking of the Columbia River (New York, NY: Hill and Wang, 2005).

Whitt, Laurelyn, *Science, Colonialism, and Indigenous Peoples: The Cultural Politics of Law and Knowledge* (New York, NY: Cambridge University Press, 2009).

Wilson, Kathleen, "Old Imperialisms and New Imperial Histories: Rethinking the History of the Present," *Radical History Review* 95 (2006), 211–234.

A New Imperial History: Culture, Identity and Modernity in Britain and the Empire, 1660–1840 (Cambridge: Cambridge University Press, 2004).

Worster, Donald, ed., *The Ends of the Earth: Perspectives on Modern Environmental History* (Cambridge: Cambridge University Press, 1989).

Index

Act Labor, 103, 121, 125, 182, *see also* legislative Acts
Actor-Network-Theory (ANT), 79–81
agency, 79
agrarian colonization, 14–18, 136–143
agrarian imaginaries, 29–31
agrarian innovation, 20
agrarian reform, 6, 17, *see also* modernity and modernization
agrarian, plantations as, 50–59, 196–197
Agricultural and Horticultural Society of India (AHSI), 20, 56
Agricultural Organization (FAO), 204
agronomy, 5–14, 19
Ahom dynasty, 42, 46
ajhar (*Lagerstrœmia reginæ*), 144
Aldrovandus, Ulysscs, 78
Almeida, Father Louis, 37
Amboyna, massacre of, 37
Amrita Bazar Patrika, The, 177, 193
ancylostomiasis (hookworm disease), 110–114, 127, 130
Andrews, C. F., 172, 176, 177, 193
anemia, 97, 110, 129
Anglo-Burmese war (1825–26), 42
Anipur garden, 169
Annual Sanitary Reports (ASR), 97, 108
anopheles mosquitoes, 14, 93, 115–116
apportionment of income taxes, 61, 65, 71–76
Arbuthnott, J. C., 149, 158
Aristotle, *Historia Animalium*, 77
Arnold, David, 19, 98, 106, 108
Asia (ship), 153–154
Assam
  agronomic growth, 200
  annexation of, 1
  disease identity, 131
  forests, 162
  government position on walkouts, 176
  growing districts, *8*
  lower, 47
  postcolonial conditions, 24
  surrounding regions and, *2*
  upper, 42, 48, 52, 56, 115
Assam Assembly, 200
Assam Bengal Railways (ABR), 104–105, 172
Assam Company
  early workings of, 47
  forest felling and, 56
  illegal contracts, 191
  land grants, 15–17, 137
  land revenue, 149
  profits, 56
Assam Forest Enquiry Committee, 164
Assam Frontier Company, 167
Assam Labour Enquiry Commission (ALEC)
  1906, 97, 129
  1921–22, 178, 185–191
Assam Land and Revenue Regulation (1886), 142
Assam plantations, 9–10, 195–200, *see also* planters
  agronomy and, 5–10
  culture and, 29–31
  early history of, 41–48
  ethno-social impact of, 4–5, 13
  images of, *10*, *139*, *166*, *184*
  local participation in, 18
  overview, 1–5
  political authority, 13
  regional politics and, 4–5
  taxes on, 50–59, 196–197
  tea yields, 87–89, *88*
  unregulated status, 195–199
  working conditions after walkouts, 195–200
Assam Railways and Trading Company (ARTC), 109, 150–152
Assam Rifles, 167
Assam Settlement Rules of 1870, 141
Assam Settlement Rules of 1886, 60

Index

Assam tea
   'discovery' of, 18, 32, 34, 42–44, 51–52
   authenticity of, 44
   botanical origins, 43
   organic, 30
Assam Valley districts, 167, 185
atmospheric theory (cholera), 105, 107, 109
Auckland, Lord, 47
authority, 13, 26, 79, 101, 131
   British Empire and, 13, 161
   ideological, 49
   improvement and, 50
   Indian Tea Association (ITA), 50–59
   natural, 50
   planters and, 13, 50–59, 64, 75, 124
   scientific expertise and, 49
   tax officers and, 73
autonomy movements, 3, 200
Azadirachta indica (neem), 204

backwardness, 44, 64, 131
'bad batch' theory, 105, 107, 127
Baden-Powell, B. H., 158, 160–164
Bahadur, Rai Radha Charan Pal, 193
Baildon, Samuel, 36, 58–59
Baitakhal, 169
Balipara sâl forests, 143
Ball, Samuel, 34
Bamber, M. Kelway, 58–59
Bangladesh, 202
Banks, Sir Joseph, 19, 41
Bantam, Java, 38
Barak Valley, 64
bari (homestead) lands, 15
Barron, T. J., 11
Bartholomaeus Anglicus, *De Proprietatibus Rerum*, 78
Barton, Gregory, 160
Barua, Ghanashyam, 65
Barua, Rosheswar, 140
basti workers, 182
Batea Banda Puk, 92
Beadon Bryant, F., 155–156, 158
Beauclerk, Captain, 151–152
Beesa, 43
beetles, 154
Behal, Rana P., 4, 22, 55, 57, 117, 120, 182
Bengal, 5, 107, 124, 146
   immigrants from, 3
   labor mortality from diseases, 111–112
Bengal Congress, 178
Bengal Forest Department, 143
Bengal Legislative Council, 193
Bengal Mahajan Sabha, 67

Bengal Tea Association, 47
Bengal, government of (GOB)
   enforcement of labor laws, 183
   forests and, 143
   labor unrest and, 169, 176, 193
   land tenure, 138–140, 142
   taxation and, 63
Bentinck, William Cavendish, 42
Bentley, C. A., 112
Berenbaum, May R., 77
beri-beri, 110, 112
Besky, Sarah, 31
Bhattacharya, Nandini, 98–99, 116
Bhuyan, S. K., 15
Bihar, 3, 63, 106–107
binaries, 31
biocolonialism, 51
bio-imperialism, 18–25
black fever. *See* kala-azar (black fever)
'Blacktown fever', 110
Bland, William & James, 47
blights, 9, 77, 90–91, 96
blister blight, 77, 81, 86
Board of Revenue, 136–138
body, 97–98, *see also* health
   health and, 101–104
   natural state of, 103–104
bohea, 54
Bombay Chronicle, The, 187, 191
Bombay Presidency, 3
Bonsall, Spencer, 56
Bordeaux spray, 12
botanical gardens, 20
botanists, 51, 66, *see also* scientific expertise
Brahmachari, U. N., 115
Brahmaputra River, 108, 154, *171*
Brahmaputra Valley, 48, 142, 186, 200
Brandis, Dietrich, 143–147, 157
British Broadcasting Company (BBC), 3
British Empire. *See also* colonialism, imperialism
   annexation of Assam, 1, 136
   as bio-political force, 23
   Assam tea and, 1
   historiography, 22–24
   non-interference, 107
   political authority of, 13
British Guiana, 3
British Medical Association, 116, 118
British West Indies, 9
Broomfield, J. H., 176
Brown, Nathan, 18, 132
Bruce, C. A., 77
Bruce, Charles, 42, 43, 45, 47, 52
Bruce, Robert, 42, 52

Index

brush fires, 90
Bryden, James L., 105
bugs, 6, 22, 199, *see also* tea pests
   use of term, 77
Burma, 146, 162
Burmese territory, 42

Cachar and Sylhet Tea Factory, *64*
Cachar district, 48, 87–89, 91, 141, 149, 153, 159, 165, 168
caffeine, 59
Calcutta, 106
Calcutta agency houses, 184, 190
Calcutta Botanical Gardens, 41, 52, 136
Calcutta High Court, 68–71
Calcutta Medical School, 111
Calcutta School of Tropical Medicine (CSTM), 111–112, 115, 116, 130
California, 89–91
Canning, Governor-General, 139, 163
Cape of Good Hope, 36
capitalism, 101
   extractive commodity, 3
   health and, 98–104
   long-term impacts of, 23–25
Carpenter, P. H., 59
Carr, E. S., 159
caste regulations, 93, 95
Catherine of Braganza, 38
Catholic missionaries, 36
Ceylon, 46, 67, 89
   coffee plantations, 11–12, 84
   tea industry, 181
Chandpur incident, 169, 175, 176, 178, 192–193
Charaideo tea estate, 149
charcoal, 133–134, 141, 143, 145–146, 156
Chargola walkout, 168, 175, 192
Charingia, Assam, *111, 113*
Charles II, 38
Charlton, Andrew, 43, 45, 52
Charter Act of 1833, 14
Chattopadhyay, Dakshinacharan, *Cha-Kar Darpan Natak*, 28, 82
Chaudhuri, J., 69
Chelmsford (Viceroy), 174
chemicals, 204, *see also* pesticides
chemists, 66, 70, *see also* scientific expertise
Chief Commissioner of Assam, 174
child labor, 197
Chin dynasty, 35
China, 67
   early history of tea in, 34–37
   East India Company and, 40–41
   tea varietals from, 22

China-Assam hybrid tea, 52–54, 67
   scientific expertise and, 44–46
Chinese tea
   compared to Assam tea, 44, 53–54, 75
   pests and, 84, 86
Chinese tea-men, 45–46, 52
cholera, 12, 28, 33, 79, 97, 100, 129–130, 199
   deaths, 109
   etiology, 119
   outbreaks, 104–110, 169–170
Christophers, S. R., 112–113
churs (riverine sandbanks), 44
Clark, J.F.M., 81
Clarke, J. J., 109–110
climate, 75, 77, 84, 87–89, *see also* microclimates
coal, 82, 150
Coal Indenture of 1881, 152
Cockrell & Co., 47
co-dependency, 134, 153–159
coercion, 13, *see also* labor exploitation
coffee plantations, 11–12
coffee rat (*Golundus ellioti*), 84
coffee rust, 84
colonialism, 79–81, *see also* British Empire, imperialism
   administrative limits of, 125
   authority and, 161
   biocolonialism, 51
   public health and, 99
Colvin, J. R., 47
comma bacillus organism, 106
commerce, culture of, 178–191
competition, 37–38, 67
Congress, 176
conservancy, 110, 113, 114, *see also* forests
Consulting Engineer to GOI for Railways, 104
contagionist theories, 105–108
contract bonuses, 191
'coolies'
   paternalist theories about, 189–190, 199
   perceptions of bodies of, 103
Cooper, W., 164
Cornwallis, 60
Cosgrave, W. A., 190
cost of living, 172–173, *see also* wages
Cotton, H.J.S., 104
Crawford, T. C., 72
Crole, David, 38, 44, 46, 54–55, 59, 70, 76
crop reduction, 181
Crop Restriction Program (CRP), 181–182, 186
Cruickshank (tea-rolling machine), 58

Index 227

CSTM. *See* Calcutta School of Tropical Medicine (CSTM)
cultivation
  homestead, 1
  income tax debates and, 66–67
culture
  definitions of, 29, 30
  of commerce, 178–191
Cuningham, J. M., 105–107
Cunningham, D. D., 112

Daily Telegraph, 178
Daisijan garden, 167
Dalhousie, Governor-General, 15, 163
Darbhanga, 106
Darjeeling tea plantations, 29–30, 86, 205
  pest management, 89–90
Darrang, 48, 183, 185, 187, 189
Das, C. R., 172, 192
Das, Rajani Kanta, 125
Daukes, F. C., 156–157
Davidson (tea drying machine), 59
De, K. C., 170
Deb, Raja Radhakanta, 42
deforestation, 157, *see also* forests
Dehing Saw Mill, 153
Demerara, 205
deodar (*Cedrus deodara*), 160
DeRenzy, Annesley Charles C., 105, 108–109, 130
desertions, 3, *see also* labor unrest
Dhekial Phukan, Anandaram, 15
Dhoedaam garden, 167, 174
Dhubri, 110
diarrhea, 97
Dibrugarh district, 94, 143, 167
Dibru-Sadiya Railway, 173
Diju forest reserve, 148
disarray, 6, 18–25, 205
disease environments, 28
diseases, human, 33, *see also* health, malaria, epidemiology, etiology, kala-azar (black fever), cholera
Donovan, Charles, 112
Doom Dooma Company, 166–167, 190
Doom Dooma district, 94, 174
Dramatic Performances Act 1876, 28
Drayton, Richard, 17, 19–20
dropsy, 97, 110
drought, 84, 87
Duars tea gardens, 112, 115
Duncan Brothers and Co., 93–94
Duplex (tea drying machine), 59
Dutch East India Company (VOC), 37

Dutch traders, 36–38
dysentery, 97, 129

earthquakes, 94
East India Company (EIC), 18
  Assam politics and, 42, 46
  military fiscalism of, 1
  monopoly, 40–41
  penal law and, 25–26
  territorial rights, 37–38
Eastern Frontier Rifles, 170
ecological indeterminacy, 20
ecology, 5–14, 194, 197
economic crisis, tea industry, 52–54, 77, 174, 179–182
economic order, 20
economics, 165, 194, *see also* profits and profiteering, capitalism, income tax debates
  agrarian, 179–191
  co-dependency of forest and tea industry, 146
  forests and, 133–135
  health and, 99–104
  health-related costs, 104, 113
  labor costs, 179
  labor mortality and, 120–132
  pests and, 77–81
  price of rice, 127, 169
  price of tea, 121
  sanitary welfare costs, 114–115
Edens, tea gardens as, 9, 18–22, 205
Edgar, J. W., 138–141
Edinburgh Botanic Gardens, 52
education, 198
Elliott, Sir Charles, 151
Empire of India and Ceylon Tea Company Limited, 172–173
Employment Review Committee, 200
Engels, Friedrich, 101
English superintendence, 75
entomology, 22, 66, 87, 90, 201–203
  imperialism and, 81
environmental history, 7, 24
Environmental Protection Agency (EPA), 204
epidemiology, 11, 19, 28, 33, 97, 130
  cholera, 105
  labor unrest and, 193
Eraligul, 169
ethnic demography, 3
etiology, 97, 104–120, 129–130
  cholera, 104–110, 119
  kala-azar (black fever), 110–115, 119
  malaria, 115–119

European agrarian superintendence, 54–59
European Economic Commission (EEC), 204
European settlement, 60
exchange-value, 71
excise duty. *See* taxes
exclusion, 101
expertise, 50, *See also* scientific expertise
export duties. *See* taxes

F. L. Engledow Commission (1936), 201–202
factory, 65, *see also* manufacturing
fair-trade tea, 29–30
Falconer, Hugh, 44
Family History of Chiang, The, 35
febrility, 110
Fee-Simple Rules (1862), 60, 139–141
fermentation, 58
fertilizers, 93, 95, 201, 203
Fiji, 3, 205
Finlay, James, 69
firing process (drying), 58–59, 69
 timber and, 133, 145
First Opium War, 1, 18, 32, 41, 49
fiscal policy, 49, *see also* income tax debates
fitness, productivity and, 103, 122, *see also* health
Fletcher, J., 69
flooding, 93, 95
flushes, 57
Food Controller, 179
food supply, 172, *see also* rice
Forest Department (FD), 6, 28, 33, 133–134, 143–164, *see also* silviculture
 Inspector-General (I-G), 143–147, 155–157
 jurisdiction, 144
 land claims, 142–143
 right-of-way and, 148–150
forests, 9, *see also* legislative Acts
 'protected', 161–162
 clearing of, 57, 197
 conservancy, 133–135, 139, 160–164
 economic aspects, 133–135
 health and, 80, 199
 private, 163
 reserved, 156–158
 unclassed state forests, 157–159
Fortune, Robert, 52
Foucault, Michel, 101
Foulis, David, 59, 118
Fraser, W. M., 94
French Guiana, 3

French travel accounts, 37
fungus, 86

Gait, Sir Edward, 14
 A History of Assam, 76
Gandhi, M. K., 6, 14, 165, 188
 non-cooperation movement, 33, 165, 169, 174, 177–178, 192–194
 on capitalism, 178
Ganges river, *171*
Gardner, Edward, 41
Gasper da Cruz, Father, 37
Gauhati (also, Guwahati), 105
Gaya, 106
Geographical Indication (GI), 29–30
George III, 19
germ theory, 110
German Law (GL), 204
Ghazipore tea estate, *83*
Ghosh, Durba, 23
ghundi pokas (green beetles), 92
Ghyabaree tea estate, 92
Gibbs (tea drying machine), 59
Giles, George M., 112–114, 130
Goalando, 169
Goalpara district, 14, 48, 144, 145
Golaghat, 86
Gordon, G. J., 42, 45, 52
Gorkhaland Movement, 29–30
grazing, 157
Great Tea Race, *39*
Griffiths, Sir Percival, 36, 40
Griffiths, William, 43–45, 52
Grove, Richard, 20, 30
Gruning, John, 148–149
Guadeloupe, 3
Guha, Amalendu, 6, 17, 55
Guha, Ramachandra, 161
Guha, Ranajit, 60
Gurkha Outrage, 172, 175, 192–193
Guthman, Julie, 29

Habib, Irfan, 40
Haffkine, Waldemar M., 106–107, 130
Hanley, Maurice, 82
Hansara tea estate, 166
Hardwar, 106
Harrison, A. John, 158
Harrison, C. J., 59
Harrison, Mark, 98, 106
hartals, 175
Harvard, F. V., 147
Haywood, H. M., 175
Hayworth (tea-rolling machine), 58

Index

hazira (quantum or volume of work), 183, 184
health, 26, 97–120, *see also* labor mortality
  body and, 101–104
  capitalism and, 102–103
  experiential, 102
  forest clearing and, 80, 199
  law and, 97–100
  laws, 120–132
  political economy of, 99–104
  productivity and, 103, 122, 124
  regulations, 103–104, 113, 120–132, 188
  wages and, 185
Heeleeka, 201
Henderson, T., 148–149
herbicides, 199
hill tribes, 42, 43, 45
Hill, H. C., 156
hoeing, 184–185, 201
Holland, Philemon, 78
homestead cultivation, 1
Honduras, 12
horticultural expertise, 49, *see also* scientific expertise
House of Commons, 198
housing, laborer, 124, 188
Hukanguri garden, 167
human management, 19, 33
hybridization, 52–54, *see also* China-Assam hybrid tea
hygiene, 26, 33, 122

ideology, 198
immigration, 120, 123, 124, 131
  disease and, 117
Imperial Entomologist of India, 92
imperialism. *See also* British Empire, colonialism
  agro-economic exploitation, 24
  bio-political, 18–25
  geopolitics, 51
  history of, 22–24
  legacy of, 3, 23–25, 33, 203–205
  trade rivalries and, 37–38
improvement, 5, 7, 13, 24, 60, 76, 198–200, *see also* modernity and modernization, progress
  authority and, 50
  disorder and, 20–22
  mechanization and, 67
  tea as, 14–18
IMS. *See* Indian Medical Service (IMS)
income tax debates, 32, 49–50, 60–76, *see also* legislative Acts
  fiscal context, 74
litigation, 68–76
indentured labor, 97
  end of, 191–192, 195
  health regulations and, 103–104, 131
  legislative acts on, 120
  migration, 3–4
  protests against, 166
  reproduction and, 117
India
  early history of tea, 39–48
  postcolonial period, 196–205
  tea exports (1929), 195
India, government of (GOI), 49
  export duties, 182
  land tenure policies, 141
  taxation debates, 62–76
  wage statistics, 184
Indian Association, 127
Indian Factories Act (IFA), 197
Indian Forester, 154
Indian Medical Service (IMS), 105, 112, 114
Indian Mining Federation, 172
Indian National Congress, 5
Indian Penal Code, 170
Indian Tea Association (ITA), 49, 130, 164
  authority of, 50–59
  fiscal crises and, 165
  labor unrest and, 175–176
  malaria investigation, 115
  pest management, 201
  recommendations on fiscal crisis, 179–181
  taxation debates and, 60–76
  wage regulations and, 197
  wages and, 184
Indian Tea Gazette, The, 85, 155
indigenous communities
  agrarian practices, 134
  disorder and improvement, 21
  forest dwellers, 160
  land tenure policies and, 136, 140
indigo, taxes on, 61–62
industrial, plantations as, 50–59, 196–197
infant mortality, 127
infection. *See also* etiology
  political trope, 165, 174
  routes of, 104–110
influenza, 95, 193
Inland Emigrants' Health Act 1889, 129
Inland Revenue Commissioners v. Ransom, 70
Inner Line Regulation of 1873, 144
insects, 200, *see also* tea pests
  use of term, 77
Integrated Pest Management (IPM), 203

irrigation, 80, 116
ITA. *See* Indian Tea Association (ITA)
Italian travel accounts, 37

Jackson (tea-rolling machine), 58
James I, 38
Jangalbari Rules, 140
Japan, early history of tea in, 36
Java, 46, 67, 89, 205
Jefferson, Thomas, 29
Jenkins, Francis, 14–18, 28, 43, 136–137
jhuming, 157
Johnson, Luttman, 183
Jorehaut Tea Company, 47
Jorhat, 86
Journal of the Agricultural and Horticultural Society of India (JAHS), 81
jungle clearing, 81, 117, *see also* forests
'jungle fever', 110, 130
jute, 66, 67

Kacharis, 1
Kadam (*Anthocephalus kadamba*), 155
kala-azar (black fever), 9, 28, 33, 79, 97, 100, *111*, 129–130, 199
  etiology, 110–115, 119
  mortality, 110
  treatment, *113*
Kamrup district, 48, 88–89, 144, 156
Kangra, 89
Karimganj district, 142, 169
Katakhal Reserve, 153
Kathmandu, Nepal, 41
Keating Saw Mill, 153
Kelman, Sander, 102
Kenrick, H. B., 63–64
Kenya, 202, 205
Kew Gardens, 19
Khamptis, 42, 43, 45
Khamtis, 52
Khilafat agitation, 169
Killing Valley Tea Company Ltd. Versus The Secretary of State for India, 68–72
Kinmond's (tea-rolling machine), 58
Kirsch, Scott, 80
knowledge economy, 6
Koch, Robert, 106, 108
Koliabar Tea Garden, 147
Kumaon-Garhwal, 44
Kumar, Prakash, 11, 62, 65, 75
Kyd, Robert, 41

labor. *See also* indentured labor
  global conditions, 192
  historians, 4
  permanent settlements, 117
  recruitment, 186
  representations of, 31–32
  skill and, 185
Labor Enquiry Commissions, 6
labor exploitation, 4, 13–14, 24, 26–27, 30
labor history, 4–5
labor impoverishment, 33, 101, 166, 178, 192, 199
  disease and, 105
  wages and, 169
labor laws. *See also* legislative Acts
  fiscal crisis and, 180
  health rules, 120–132
  on sanitary welfare, 114
  penal provisions, 117, 121–122, 127, 128, 129, 190–191
  wages and, 182–191
labor migration. *See* immigration
labor mortality, 3, 9, 97–120, *see also* health
  administrative data on, 125
  Ceylon coffee plantations, 12
  diseases and, 33, 79, 110–111
  economic aspects, 120–132
  law and, 120–132
  living conditions and, 22
  rates, 6, 27, 107, 118–121, 123, 125–129, *126, 128*
labor unions, 165, 192, 198, 199
labor unrest, 13, 19, 165–166, *see also* walkouts, protests
  1920–21, 166–172
  causes of, 166, 172–179
  government neutrality and, 176
  politics and, 176
Lackwah Tea estate, 148
Lakhimpur district, 48, 87–89, 118, 158, 159, 185, 190
land grants, 17, 28, 136–141
land revenue, 28, 60, 142
land tenure policies, 60, 136–143
landscape manipulation, 33
landslides, 94
Lansdowne (Viceroy), 157
Latour, Bruno, 20–21, 79
latrines, 110, 114, 127, 197
Law, John, 79
lawlessness, 21–22, 27, 190, 194, 195–199, *see also* disarray
laws. *See also* labor laws, legislative Acts
  disorder and, 25–28
  public health and, 97–100
Leasehold Rules of 1876, 60, 141, 145, 151

Ledo, 150
Lefroy, Harold Maxwell, 92
legislative Acts, 6
  Labor Acts
    Act XIII of 1859, 121, 122, 190–191
    Act III of 1863, 120, 123
    Act VI of 1865, 120, 123, 182–183
    Act VII of 1873, 120
    Act I of 1882, 121, 124–125, 126–128, 130, 131, 183
    Act VI of 1901, 122, 129, 188, 190, 198
    Act XI of 1908, 122
    Act VIII of 1915, 188
    Act XII of 1920, 191
  Income Tax Acts
    Act II of 1886, 57, 60–63, 65, 75–76
    Income Tax Act VII of 1918, 182
    Income Tax (Revised) Act XI of 1922, 72
  Forest Acts
    Forest Act VII of 1865, 133, 147, 161
    Forest Act VII of 1878, 133, 144–145, 156–157, 161–162
    Assam Forest Regulation VII of 1891, 133, 145, 149, 156–159, 164
    Indian Forest Act XVI of 1927, 162
  Wasteland Acts
    Wasteland Rules of 1838, 60, 137
    Wasteland Rules of 1854, 150
    Wasteland (Leasehold) Rules of 1876, 60, 141, 145, 151
Leishman, William, 112
lime, 90, 201
Liners Conference (1920), 182
Linnaean taxonomy, 55, 78
Linné, Carl, *Systema Naturæ*, 78
Linschooten, Jan Hugo van, 37
  Linschooten's Travels, 38
locked-up lands, 145
Longai, 169
Longai Valley, 142
Longley, P.R.H., 92
looper caterpillars, 77
Lower Jiri Reserve, 153
Lu Yu, 36
Lum Qua (or Lumqua), 42, 45, 46
Lumding, 104–105
Lyall, C J, 87

M'Clelland, John, 52
M'Cosh, John, 48
Macao, 36
Macaulay, Thomas Babington, 25–26

Mackenzie, 91
Madras, 162
Makai (*Shorea pennicellata*), 155
Malacca, 36
malaria, 9, 12, 28, 33, 79, 97, 100, 110, 129–130, 197, 199
  anopheles and, 93
  etiology, 115–119
  prevention of, 116–120
malaria cachexia, 110–111
Malaria Committee of the Royal Society to India, 112
Malaya, 3
malguzari forests, 163
malnutrition. *See* starvation, nutrition
Mandelslo, Johan Albrecht von, 40
Manipuri tea, 86
Mann, Gustav, 143, 146, 147, 151
Mann, Harold H., 59, 86, 94, 201
manufacturing, 57
  taxation and, 61–76
Margherita, 150
Martinique, 3
Marwari kayas, 174
Marx, Karl, 71
Marxism, 101
Marxist historiography, 13
Masters, J. W., 56
Mauritius, 3, 205
McCabe-Dallas, Alfred, 112
McClelland, J., 43–45
McKercher, W. G., 72–73
McPherson, J., 170
McRae, W., 86
mechanical know-how, 75
Meckla Nadi Saw Mills, 158
medical arrangements, 105, 115, *see also* health
  inspection of labor, 123–124, 127
medical men, 103, 107
medicine, scholarship on, 98–99
Merchant, Carolyn, 18, 94
Meyer, Sir William, 66
microbes, 200
microclimates, 85, 202
middle-class, Assamese, 5, 15
Mildmay, Walter J., 148
Milligan, J. A., 72
Mills, A.J.M. Moffatt, 14–18, 137–138
Milroy, A. J., 152
mineral resources, 150
Minimum Residue Levels (MRL), 204
minimum wage, 129, 183, 185, *see also* wages
Misings, 1

232   Index

Mitchell, Don, 80
Mitchell, Timothy, 12, 205
Mitter, B. C., 71
modernity and modernization, 6, 13, 17, 19–20, *see also* improvement, progress
   agrarian, 49
   colonial, 11
   human/nonhuman divide and, 20–21
   income tax debates and, 66
   tea industry and, 64–68, 76
   tea manufacturing and, 76
Mohapatra, Prabhu P., 9, 117, 120
mole cricket, 77
Money, Edward, 53–58, 59, 70, 75–76
monocultures, 51, 80, 85, 95
   pests and bugs, 22
   unsustainability of, 12
monsoons, 82, 87, 182
Mookerjee, Sir Asutosh, 69–71
morbidity, 100, 103, 124, 131
Morning Post, 174
mortality. *See* labor mortality, infant mortality
Mughal Empire, 39–40
Mullane, J., 127
Muriel, C. E., 159
Muttock tea tract, 42, 77
mycologists, 201

Nag, Hardayal, 170
Nambor forest, 143, 150
nationalism, 165, 192–194
natural calamities, 94–95
nature, as ecosystem context, 9
Navigation Act of 1651, 37
neem tree (*Azadirachta indica*), 92
neerikh, 185
   (rate for task-work), 183
Nelson (tea-rolling machine), 58
Neo-Han dynasty, 34
Nepal, 5, 41
Netherlands, 37, *see also* Dutch traders
New Delhi, 200
New York Times, The, 3
Non-Act Labor, 121, 125
non-regulation, 21–25, *see also* disarray, lawlessness
North Lakhimpur, 86, 94, 168
North Sylhet district, 185
Northern Sung dynasty, 35
Northern Whig, 177
Nowgong (also, Nagaon, Nagoan), 48, 68, 87, 118, 148, 190
nutrition, 26, 118, 122, 127

O'Callaghan, Mr., 168
Old Assam Rules (1854), 60, 138, 163
Olmstead, Alan L., 12
opium, 17, 41
organic farming, 29–30
Orissa (now Odisha), 63
Ovington, John, *Voyage to Suratt*, 40
oxidation, 58

P.G. ('practical gardener'), 55
Pabhajan tea estate, 167, 174
Pal, Bipin Chandra, 172
Panama Canal, 12
Pande, Radha Krishna, 169
Panitola tea estate, 94, 189
Paramaribo, 4
Parsons, Talcott, 102
pastoral visions, 29–31
paternalistic theories, 189–190, 199
Pathalipam estate, 168
pathogen ecologies, 196
Pattle, James, 42
Paul, Sir G. C., 61, 70
Peal, Samuel E., 81–85, 90–91, 93, 155, 202–203
peasants, 17, 136
   agriculture, 19
   forest conservancy and, 160
   protests, 134, 163
   settlements, 153, 157
Pegu (Burma), 144
penal code (IPC) of 1837, 25
Penang, 45
Pepys, Samuel, 38
Permanent Settlement proclamation, 60
pest ecology, 11
pest management, 79, 200–205
   Bordeaux spray, 12
   knowledge networks, 89–96
   labor and, 84, 90
pesticides, 22, 199, 202–204
Petroff, Ivan, 37
petroleum, 82, 150
Pettenkofer, Max von, 109
Phillimore, Sir Robert, 153
phytochemicals, 204
Pitt the Younger, 19
plantation science, 11–12
plantation-as-family discourse, 30
planter violence, 3, 26, 28, 122, 196, 197
Planter's Raj, 5, 13, 18, 195
planter-as-farmer, 30
planters. *See also* Indian Tea Association (ITA), Assam plantations
   as monopoly capitalists, 142

Index 233

Assamese, 140
  authority of, 50–59, 75
  cholera etiology and, 107
  health regulations and, 103
  land grants, paternalistic favoritism and, 136
  legitimate authority and, 64
  taxonomy analogy, 55
Pliny the Elder, *Historia Naturalis*, 77
plucking, 69, 181–182, 185, 203
politics. *See also* nationalism, protests
  global, 192
Portuguese traders, 36–37
postcolonial period, 196–205
postcolonial theory, 23
poststructuralism, 23
poverty. *See* labor impoverishment
power, 5, 11, 13, 23, 26, 49, 64, 79, 80, 131, 164, 197, 200, *see also* authority
  hierarchies of, 92
  of tax officers, 73
power-relations, 196
pregnancy, 124
primitivism, 43, 69, 131, 198
Prinsep, W., 47
Privy Council, 71
productivity, health and, 103, 122, 124
profits and profiteering. *See also* economics
  disease and, 110, 114
  diseases and, 199
  pests and, 79
  quantity and, 179
progress, 5, 13, 20, 76, 200, *see also* improvement, modernity and modernization
proletarianization, 32
prophylaxis, 130
protests, 3, 31, 33, 165–179
  1920–21, 166–172
  by peasants, 134, 163
  causes of, 176, 192–194
  pest management and, 93
  wages and, 179–191
pruning, 57, 86, 90, 203
public health. *See also* health
  capitalist profits and, 98–104
  colonial power, 99
  scholarship on, 98–99
  vaccination and, 107
public meetings, restrictions on, 170
Public Works Department (PWD), 143

quinine, 113, 115–116, *119*

race prejudice, 126
racial ideologies, 75
racial typologies, 45
Raidang estate, 167
rainfall, 78, 82, 86, 87–89
  pests and, 85
Ramsay, G. C., 118, 199
Ramusio, Giambattista, *Navigatione et Viaggi*, 36
Rangarajan, Mahesh, 24, 162, 163
Rappaport, Erika, 38
Ratabari garden, 169
RCLI, 198
rebellion
  by Singphos, 46
recovery, 19
red spider (tea mite, *Tetranychus bioculatus*), 28, 77, 84–87, 88, 91, 202
refinement
  ideology of, 75
regulation
  profiteering and, 188
Reid, D., 149–150
Report of the Indian Industrial Commission (RIIC), 67–68
reproduction rates, 117
resistance, 19, *see also* protests
resource management, 133–135, *see also* Forest Department (FD)
respiratory disease, 97
retrenchments, 186
return-to-nature, 30
Réunion, 3
Revenue Authorities, 72
Revenue Secretary, GOI, 156
Rhode, Paul W., 12
Rhodes, Father Alexander de, *Voyages et Missions Apostoliques*, 37
Ribbentrop, Berthold, 157
rice
  cultivation, 15, 117–118
  for workers, 167–168
  mechanization processes and, 67
  price of, 127, 169
  provided by planters, 188
riots, 3, 32
Rogers, Sir Leonard, 111–112, 115
rolling and rolling machines, 58–59, 69
Rosenberg, Charles, 130
Ross, Sir Ronald, 112, 115
Roy, Tirthankar, *74*
Royal Commission on Agriculture in India (RCAI), Evidence Taken In Assam, Vol. V, 72

234 Index

Royal Commission on Labor in India (RCLI) 1930, 196–199
Royle, J. Forbes, 44
rubber extraction, 157
ruination, sociology of, 203
Rupacherra tea garden, 149
rupit (wet-rice cultivating) lands, 15
Russian tea consumption, 179
Russian travel accounts, 37
ryots, 136, 137, 138, 141, 147, 149

Sadiya, 43, 46, *119*
'saheb's disease'
Saikhowa, 45
Saikia, Arupjyoti, 143, 150, 152, 153, 157
sâl (*Shorea robusta*), 142, 143, 148–149, 156, 160
Salona Tea Company, 148
salt, 90
sam (*Artocarpus chaplasha*), 144, 155
Samdang garden, 167
sanitarians, 103
sanitary commissioners, 107–110
sanitary conditions, 105
  cholera and, 109
  kala-azar and, 114
  profits and, 121–122
Sanitary Department of GOI, 105
sanitary measures, 26, 122, 124–125, 128, 131, 188
'sarkari bemari' (British Government disease), 110
saw mills, 154
Schiebinger, Londa, 44
Schlich, Wilhelm, 143
scientific expertise, 11, 22, 51, 66
  China-Assam hybrid tea and, 44–46
  ideological authority and, 49
  pest management, 29, 78, 89–96, 201–204
  practical know-how and, 46
  vetting Assam tea, 43–44
scientific propriety, 110
scientific rationality, 19
scientific triumphalism, 32
Scinde (ship), 97
Scott, Beatrix, 91
Scott, David, 14, 42
secessionism, 3, 200
Sen, Ram Comul, 42
Sewell, Henry, 18
sexual violence, 26
shade, 81, 87
Sharma, Jayeeta, 15
  Empire's Garden, 24

Shen Nung, *Pen ts'ao*, 34
Shibpur, 41
Short, Thomas, 38
Sialkot, 106
Sibsagar district, 48, 81, 82, 115, 144, 145, 148, 168, 185
Sigatoka fungus, 12
Sigmond, G. G., 35, 48
silviculture, 28, 33, 134–135, 139, *see also* forests
Sim, G. G., 72
Simul tree (*Bombax malabaricum*), 154–155, 158
Singapore, 45
Singha, Purandhar, 42
Singha, Radhika, 26
Singha, Raja Purandhar, 46
Singpho tea tract, 77
Singphos, 42, 43, 45–46, 51
Sirocco (tea drying machine), 59
site selection, 116, 119
Sivaramakrishnan, K., 135, 160
slavery, 4
Smythies, A., 154
social agency, 79
social inequality, 101
soil conditions, cholera and, 109
Sola forest reserve, 148
Soluri, John, 10
sources, elite, 32
Spivak, Gayatri, 31
Spry, H. E., 72
Sri Lanka, 202, 205, *see also* Ceylon
standard of living, 198
starvation, 169–170, 176, 193, *see also* nutrition
Stoler, Ann, 204
Straits Settlement, 45
Strickland, C., 116–119, 130
strikes, 166, 175
Stuart, C. P. Cohen, 36
subaltern consciousness, 31
subsistence wages, 186
sugar industries, 63
Sung dynasty, 36
Surinam, 3–4
Surma Valley, 48, 140, 142, 158, 165, 168, 172, 175, 176, 181
  tea quality, 185
  wages, 186
Sutter, Paul, 12
Swargadeos (Ahom god-king), 15
Sylhet district, 14, 48, 118, 165, 168–169, 177

Index

Taiwan, 202
Talish, Shihabuddin, *Fathiya-i 'Ibriya*, 40
taxes, 14, 17, 196
  agriculture and, 60–76
  apportionment, 61, 65, 71–76
  evasion of, 22
  excise duty on opium, 17
  excise duty on tea, 38
  export duty, 72, 182
  import duty on tea boxes, 72
  on petrol, 72
  tea-box royalty, 159
taxonomy, 78
tea. *See also* Chinese tea, China-Assam hybrid tea, Assam tea
  as Britain's national drink, 38
  as leaf, 50, 74–75
  botanical origins, 36
  circulation and exchange, 36–39
  cultivation process, 56–57
  early history of, 34–44
  indigenous, as wild, 51
  introduction to Europe, 36–37
  manufacturing process, 57–59
  medicinal aspects, 35, 37
  production (yields), 87–89, *88*, 168, 179–182
  quality, 181, 185
tea aphis, 77, 82–84
'Tea as Progress', *16*
Tea Board, 29
Tea Committee, 17, 42–45, 52
  agrarian colonization and, 136
Tea District Labour Association, 170
tea experts, 66
tea gardens. *See* Assam plantations
tea grub, 92
tea industry
  after World War I, 195–200
  agro-economic problems, 77
  crop output guidelines, 181–182
  downturn in, 121
  during World War I, 179
  economic crisis, 52–54, 174, 179–182
  market rivalry, 76
  modernity and, 66–67, 76
  production levels, 165
  recession, 140–141
tea juice, 84
tea mite. *See* red spider (tea mite, *Tetranychus bioculatus*)
tea mosquito bug (*Helopeltis theivora*), 28, 77, *78*, 81–84, *83*, 90, 201–202

tea pests, 6, 9, 22, 28–29, 32, 77–96, 197, 199, 200–205, *see also* pest management
  economic aspects, 77–81
  species in northeast India, 85
tea plant (*Camellia drupifera*), 42
tea plant (*Camellia sinensis var. assamica*), 1
tea plant (*Camellia sinensis*), 34, 64, 74
tea-boxes, 133–134, 141, 145–146, 153–159
teak (*Tectona grandis*), 133, 143, 160
tea-rolling machines, 58
technical know-how, 75
Terai region, 201
Tezpur division, 143
Thea Bohea, 46
Thiselton-Dyer, Sir William Turner, 55
Thompson, E. P., 31
Thompson, Edgar, 75
Thompson, W. J. & H., 47
thrips (*Scirtothrips dorsalis*), 77, 86–87
ticca (piece-work), 168, 176, 183, 185, 187, 188
Tikak, 150
timber, 142, *see also* forests
  local use of, 144
  marketability of, 156–157
  revenue, 150–151
Times (London), 179
Tinker, Hugh, 4
titapani, 92
tobacco, 7
Tocklai, 201
trade rivalry, 37–38
transplanting, 57
travelers, 36, 51
Travers & Sons, 47
tripartite moral economy, 30–31
Twining & Co., 47

Uekötter, Frank, 9
Ukers, William, 34, 46
unit system, 183
United States
  agricultural innovation, 12
  organic agriculture, 29
United States Patent office, 56
unkemptness, 19, *see also* disarray
unrest. *See* labor unrest
unruliness, 21
Upper Assam Tea Company, 143
Upper Burma Regulation of 1887, 157
use-value, 71
Uttarakhand, 44

vaccination, 107, 125, 127, 130
Venesta box, 155
Vetch, H., 15
Victoria (tea drying machine), 59
Vidyaratna, Ramkumar, *Kuli Kahini*, 82
violence, 13, *see also* planter violence, sexual violence
  against laborers, 170

wages, 9, 27, 117, 121, 174
  differential wage structure, 181–191
  labor demands, 172–179
  labor recruitment and, 186
  labor unrest and, 194
  regulation of, 197
  statistics, calculation of, 187
walkouts, 3, 32, 165
  1920–21, 166–172
  causes of, 166
  government response to, 175–177
  wages and, 176, 186–191
Wallich, Nathaniel, 42–43, 52, 136
Ward, Sir William, 150, 152
Wasteland Rules of 1838, 60, 137
Wasteland Rules of 1854, 150
Wasteland Rules of 1876, 151
wastelands, 28, 93, 133
  availability of, 152
  colonization of, 136–143
  forest cover, 147
water supply, 108–110, 113, 114, 124
water-borne theory of cholera, 105, 107
Watson, Sir Malcolm, 119
Watt, Sir George, 90, 203
weather. *See* climate
Webb, Charles, 82

weeds, 200, 203
welfare provisions, as concessions, 188–190
wellness, 103, *see also* health
West Bengal, 200
Wheeler, Sir Henry, 175–176
  Report, 177
White violence, 25–26, *see also* planter violence
White, J. Berry, 46, 54
Whitt, Laurelyn, 51
wilderness, 7
wildfires, 95
withering, 57–58, 69
women, 30
  reproduction, 117, 124
  wages, 183, 185
wood, use in tea industry, 153–159, *see also* timber, tea-boxes
woodlands, 28
Wood-Mason, James, 82–87, 90, 203
Woodroffe and Evans (W & E), 61–62, 70
worker solidarity, 192–194
working conditions, 57, *see also* labor impoverishment, labor laws, wages, protests, sanitary conditions
  protests against, 172–179
World Health Organization (WHO), 204
Wyndham, W. Y., 201

Yalysheff, Boornash, 37
Young, T. C. McCombie, 114–115
Yunnan, 42

zamindars, 157, 160, 163
Zimbabwe, 202